THE HANDLEY PAGE HASTINGS AND HERMES STORY

by

Victor F. Bingham
T. ENG (CEI), AMRAeS, AFSLAET

Illustrations
by

Lyndon Jones
MCSD

HANDLEY PAGE HASTINGS AND HERMES

First published 1998
by GMS Enterprises
67 Pyhill, Bretton, Peterborough,
England PE3 8QQ
Tel and Fax (01733) 265123
EMail: GMSAVIATIONBOOKS@ Compuserve.com

ISBN: 1 870384 63 6

Copyright: Victor Bingham

All rights reserved.
No part of this publication may be reproduced,
stored in a retrieval system, or transmitted,
in any form or by any means, electronic, mechanical,
photocopying, recording or otherwise
without the prior permission of the publishers.

Printed and bound for GMS Enterprises

Contents

Acknowledgements		iv
Preface		v
Abbreviations		vii
Bibliography		viii
Chapter One	Air transport development	1
Chapter Two	H.P. 67 Hastings development	8
Chapter Three	Production	17
Chapter Four	Testing and analysis	21
Chapter Five	The Hastings aircraft	27
Chapter Six	Service use of the Hastings	37
Chapter Seven	Hermes aircraft development	69
Chapter Eight	Hermes in servce	82
Appendix One	Basic data on Hastings aircraft	96
Appendix Two	Basic history of each Hastings aircraft	97
Appendix Three	Basic data on Hermes aircraft	102
Appendix Four	Basic history of each Hermes aircraft	103

Acknowledgements

This book is the result of a great deal of research and the perusal of Handley Page documentation, records at the Public Record Office and RAE Farnborough. Many people have helped and encouraged me, too numerous to list, including many Hastings, Hermes and Handley Page personnel. Many have kindly loaned me documentation and photographs. To all I tender my grateful thanks, especially to the following, not in preferential order or alphabetically.

S/Ldr K.R.Jackson MBE.AFC and Warrant Officer P.W.Porter BEM RAF for photographs and notes on the Hastings and its operating. A.G 'Jack' Knivett ex-Handley Page technical representative for photographs and technical notes. Richard Searle, Denis Goode and staff RAE Farnborough library/archives. Brian Kervell curator of RAE Farnborough Museum. Matt George and Tim Calloway of RAF Museum Hendon. Staff of the Public Record Office Kew. Brian Gardner of British Caledonian. Bob Bullock, Peter Caines, Captain Cope, Brian Hutchinson, Harry H.Pusey, L.F.Painter, Bob Stanbridge, T.A.Tonkinson and Captain A.R.Thompson DFC of BOAC. Ken Lyons ex-Handley Page and RAE Structures. Cedric Vernon ex-Handley Page Chief Aerodynamicist and ARB. Harry Fraser Mitchell, Alan Dowsett, E.N.Brailsford and Peter Cronbach, all ex-Handley Page. Bernard Wheeley ex-Airwork and BEA. Maurice Patterson, Ken Heseltine and A.N.Mitchell ex-RAF. Derek Reed of Yorkshire Air Museum. S/Ldr J.Sabin and F/Lt G.Heath RAF.

Special thanks to S/Ldr 'Jacko' Jackson, A.G.Knivett and John Strickland for reviewing the manuscript and offering comments.

Finally a thank you to my wife, who some way or other always managed to tolerate my typing and files, as well as my frequent disappearances on aircraft research.

Preface

This is the story of two Handley Page transport aircraft, and covers their design, development, testing and service. So if you think that the story of transport aircraft is uninteresting we challenge you to read on. For this is the story of two aircraft that did not get overwhelming support from Government funds. This is the story that illustrates the changing requirements and muddle post-war of the 'powers-that-be' in the RAF, MAP and BOAC. It illustrates the post-war problems that British aircraft manufacturers had to contend with and overcome, when neither the RAF nor the designated national carrier appeared to know what they wanted.

The old established firm of Handley Page, having produced the Halifax in many forms to serve the RAF in many roles, entered the post-war years with two projects, the HP67 Hastings military transport and the HP68 Hermes civil airliner. Both were conceived during the latter part of the war and developed under the old concept of building and flying an aircraft shell to prove the basic concept and services, followed later by the fitting out of the aircraft for its roles. Both were adaptations of the Halifax airframe, and both were, under the pressure of the time and demands for improvements or changes of requirements, designed away from their original concept.

As will be seen in the text, initially the Hastings and Hermes were so closely connected in their design that their story is closely entwined, the Hermes connection in the early years continually flitting in and out of the Hastings story - thus continually indicating the lack of awareness of the loss of time, as well as the amount of indecision. Also illustrated is Handley Page's commitment to providing the Royal Air Force with a large transport aircraft. The firm at that time still having the same team that had designed, developed and produced a long line of effective aircraft for the Royal Air Force. The same team that would design and produce the last and the best of the V-bombers, the same team that would be scattered when the firm was decimated by Government intervention.

Both aircraft had trials and tribulations, both served or entered war zones, both were well-built aircraft in the Handley Page tradition, both were interesting aircraft. At that particular time at the end of the war, the Douglas DC4 was considered to be the yardstick of performance of a transport aircraft, yet the Hastings - in spite of its adaptation and redesign - was to prove as good a heavy hauler, but with a better cruise and top speed. Between 1948 and 1967 the Hastings was the major transport aircraft of RAF Transport Command, and for over 25 years served in one role or other in the RAF, RNZAF and the Ministry of Supply and its successors into the 1970s.

In that period it had its baptism on the Berlin Air Lift, was 'under fire' during the Suez War and the confrontation with Indonesia. It transported men and materials all over the world; east to Hong-Kong, Australia and Christmas Island, and west to the Americas - not to mention Greenland. It operated on airborne support carrying loads internally and externally; on meteorological duties, VIP transport, radar training and as a research aircraft - as we have said, an interesting aircraft having interesting roles.

The Hermes was developed from its original 'tail dragger' configuration to become the Hermes IV and V with a comprehensive pressurisation system and a tricycle undercarriage. Handley Page's achievement both technically and commercially in respect of this final development away from the Halifax family, was born out of a real effort to get out of the tight strait-jacket of wartime military restrictions and to compete in the civil market. That this expenditure of effort and finance failed to bring in large orders for the Hermes IV does not reflect on the company and employees, for it was competitively priced and engineered. In the design and development of the systems and equipment for the Hermes IV was born the design and testing of equipment for the company's Victor bomber - ploughshares to swords.

The Hermes IV went into service with BOAC after delays, both technical and to lack of decision, only to be ousted by jet transport within two years. It then gave excellent and reliable service with the Independents for a number of years, both at home and overseas; like the Hastings, proving the soundness and strength of its construction - another Handley Page aircraft.

I make no apologies for considering the Hastings and Hermes as interesting aircraft, worthy of consideration as leading transport aircraft of their time. That the aviation press has seen fit to basically ignore them is a loss to aviation enthusiasts. So I dedicate this book to all Handley Page personnel who designed, produced and tested the Hastings and Hermes, and to those who flew and serviced these aircraft. Opinions and comments in this book are my own, unless otherwise stated.

Victor F. Bingham.
January 1998

HANDLEY PAGE HASTINGS AND HERMES
Abbreviations

A&AEE	Aeroplane and Armament Experimental Establishment.
AFC	Air Force Cross.
AFEE	Airborne Forces Experimental Establishment.
AGL	Above ground level.
AID	Aeronautical Inspection Directorate.
AMSO	Air Member for Supply and Organisation.
ANS	Air Navigation School.
AOC	Air Officer Commanding.
AP970	Air Publication 970. (Manual on structures and tests).
ARB	Air Registration Board (civil aviation).
ATA	Air Transport Association (USA regulatory body).
ATC	Air Traffic Control.
AVM	Air Vice Marshal.
BOAC	British Overseas Airways Corporation.
BCBS	Bomber Command Bombing School.
CAB	Civil Aviation Board.
CAR	Civil Aviation Requirements
C of A	Certificate of Airworthiness.
CG	Centre of Gravity.
CGCA	Control General of Civil Aviation.
CRD	Controller of Research and Development.
CSU	Constant speed unit (Propeller).
CSE	Central Signals Establishment.
DBR	Damaged beyond repair.
DD	Deputy Director
DGRD	Director General of Research and Development.
DOR	Director of Requirements.
DTD	Directorate(or Director) of Technical Development.
DV	Direct vision (openable window for clearer vision).
EAS	Estimated air speed.
EFS	Empire Flying School.
ETPS	Empire Test Pilots School.
FECS	Far East Communications Squadron.
'HP'	Sir Frederick Handley Page.
HF	High Frequency.
HM	Her Majesty
hp	Horse Power
HPA	Handley Page Association
H2S	Code name for radar navigation and bombing aid.
IAS	Indicated air speed.
ICAN	International Convention on Air Navigation.
IFF	Interrogation Friend or Foe.
IRIS	International radio installations and systems.
MAP	Ministry of Aircraft Production.
M of A	Ministry of Aviation.
MBE	Member of the British Empire.
MCA	Ministry of Civil Aviation.
MEA	Middle East Airlines.
MECS	Middle East Communications Squadron.
Met	Meteorological.
MF	Medium frequeney.
MoS	Ministry of Supply.
NACA	National Advisory
NRV	Non-return-valve.
OCU	Operations Conversion Unit.
OTU	Operational Training Unit.
PTS	Parachute Training School.
PTU	Parachute Training Unit.
QANTAS	Queensland & Northerm Territories Air Services (Australia).
QNE	'Q' code for altimeter millibar setting at airport of arrival.
RAE	Royal Aircraft Establishment.
RAFFC	Royal Air Force Flying College.
RCAF	Royal Canadian Air Force.
RDA	Department in Air Ministry covering R&D.
RDAU	Department in Air Ministry covering R&D.
RDAP	Department in Air Ministry covering R&D.
R&D	Research and Development.
RNZAF	Royal New Zealand Air Force.
RRE	Radar Research Establishment.
R/t	Radio telephony.
RTO	Resident technical officer (MoS teehnical representative).
SAC	Strategie Air Command.
SAS	Special Air Service.
SBAC	Society of British Aircraft Constructors.
SCBS	Strike Command Bombing School.
SOC	Struck off charge.
SNCO	Senior non-commissioned officer.
TCA	Trans-Canadian Airlines
TCASF	Transport Command Air Support Flight.
TCDF	Transport Command Development Flight.
TCDU	Transport Command Development Unit.
TKS	Name of de-icing equipment company.
TRE	Telecommunications Research Establishment.
UHF	Ultra high frequency.
UK	United Kingdom.
USAF	United States Air Force.
VDF	Variable direction finding.
VHF	Very high frequency.
VIP	Very important person.
WEE	Winterisation Experimental Establishment.
WFU	Withdrawn from use.
WM	Weak mixture.(refers to fuel metering).

Bibliography

Literature on the Hastings and Hermes aircraft mainly centres on the official files and reports held at the Public Record Office Kew, the Royal Aircraft Establishment Farnborough, and the Aeroplane and Armament Experimental Establishment Boscombe Down. This is rather surprising as the Hastings and Hermes were widely used over a number of years, all over the world. The following list of literature may thus be of interest to the reader.

'Halifax - Second to None' by V.F.Bingham, published by Airlife Publishing.

'Handley Page aircraft since 1907' by C.H.Barnes, published by Putnam.

'Rescue below Zero' by Ian Mackersey, published by Robert Hale.

'British Civil Aircraft 1919 -1959' Vol 2 by A. J. Jackson, published by Putnam.

'Test Pilots' by Don Middleton, published by Willow Books.

'Handley Page' an aircraft album by Donald C.Clayton, published by Ian Allan.

'The Aeroplane' May 24. 1946.

'Air Clues' July 1985.

'Aircraft Production' February, March, May and June 1949.

'Flypast' July 1986.

Chapter One
Air Transport Development

Transport aircraft trends.
The pre-1939 British air transport scene was noticeable for its lack of British designed modern large transport aircraft, military or civil. The most modern being the Armstrong Whitworth Ensign and the Handley Page Harrow, the latter aircraft perpetuating the bomber transport type so beloved of the RAF since the 1920s. The De Havilland Albatross was a modern design, but could hardly be classed as a large aircraft, and its role was mainly as a fast mailplane or a short range passenger aircraft.

To rectify the civil situation, the Ministry issued two specifications - possibly more in hope than expectation! The 14/38 for a long range civil passenger aircraft and 15/38 for a short/medium haul civil passenger type. A number of firms tendered to the specifications and the choice eventually settled on the following:

	Fairey FC1	**Short S32**
Specification	15/38	14/38
No. of prototypes	one	three
No. of engines	four	four
U/C type	tricycle	tricycle/tailwheel
Wingspan	105 feet	127 feet 6 inches
Wing area	1300 sq feet	2020 sq feet
Length	83 feet	90 feet 9 inches
Est. AUW @ T/O	45,000 lbs	71,000 lbs
Est. max. speed	275 mph	275 mph
Est. range	1700 miles	3400 miles

With the commencement of war on the 3 September 1939, designs and developments not connected with the war machine were phased out, the FC1 being cancelled in October 1939 and the S32 being abandoned in May 1940.

During the years 1939-45 the Allied air transport scene was dominated by the well established Douglas DC3, followed later by the DC4 (known as the Skymaster), which first flew on 14 February 1942 and was quickly adopted by the U.S military as the C54, with 1315 being built before it was superseded in production by the DC6.

In the RAF at the start of the war the Handley Page Harrow and Bristol Bombay, with support from the commandeered civil aircraft, formed the air transport contingent. By the mid war years the RAF Air Staff issued their 'Transport A' and 'Transport B' specifications, for which Handley Page tendered and produced their Halifax C3, C6, C7 and C8 series of aircraft. At this period in the airborne support role were operated relatively small numbers of the Albemarle, Halifax and Stirling aircraft, that supplemented the Dakotas (DC3) transports.

Fairey Aviation, having originally tendered to the pre-war transport specifications, was now a 'daughter' firm manufacturing the Halifax bomber, and saw in this airframe the basis of a transport aircraft. So during 1941-42 began design studies for converting this aircraft into a tricycle undercarriaged low-wing airliner/transport. Unfortunately, by the time that the Brabazon Committee had been formed to look into post war air transport the project appears to have been laid aside - or suppressed, and so remained only a paper project (artist's impression Drawing 1).

Avro's were not to be left out of the transport side of aviation and in 1942 proposed to the Air Staff a project where a large boxlike fuselage would be mated to Lancaster mainplanes, tail unit and undercarriage. As opposed to Handley Page's

Table 1. Comparison of basic details.

	Avro York	**Tudor 1**	**Douglas DC4**
Wingspan	102 ft	120 ft	117.5 ft
Wing area, sq. ft.	1205	1421	1460
Length	78.5 ft	93 ft 11	93 ft 11
Empty weight, lbs.	38,597	48,217	37,000
AUW at take off, lbs.	71,000	70,000	73,000
Wing loading, lb/sq. ft.	50.2	49.25	42.5
Power loading, lb/h.p.	12.7	10.42	12.00
Cruising speed	233 mph	230 mph	203 mph
Maximum speed	298 mph	346 mph	265 mph
Service ceiling, ft.	26,000	28,800	19,000
Range, miles	2700	2400	2140
Payload, lbs.	6420	6262	14200
Take off to clear 50 ft.	1880 yards	950 yards	950 yards

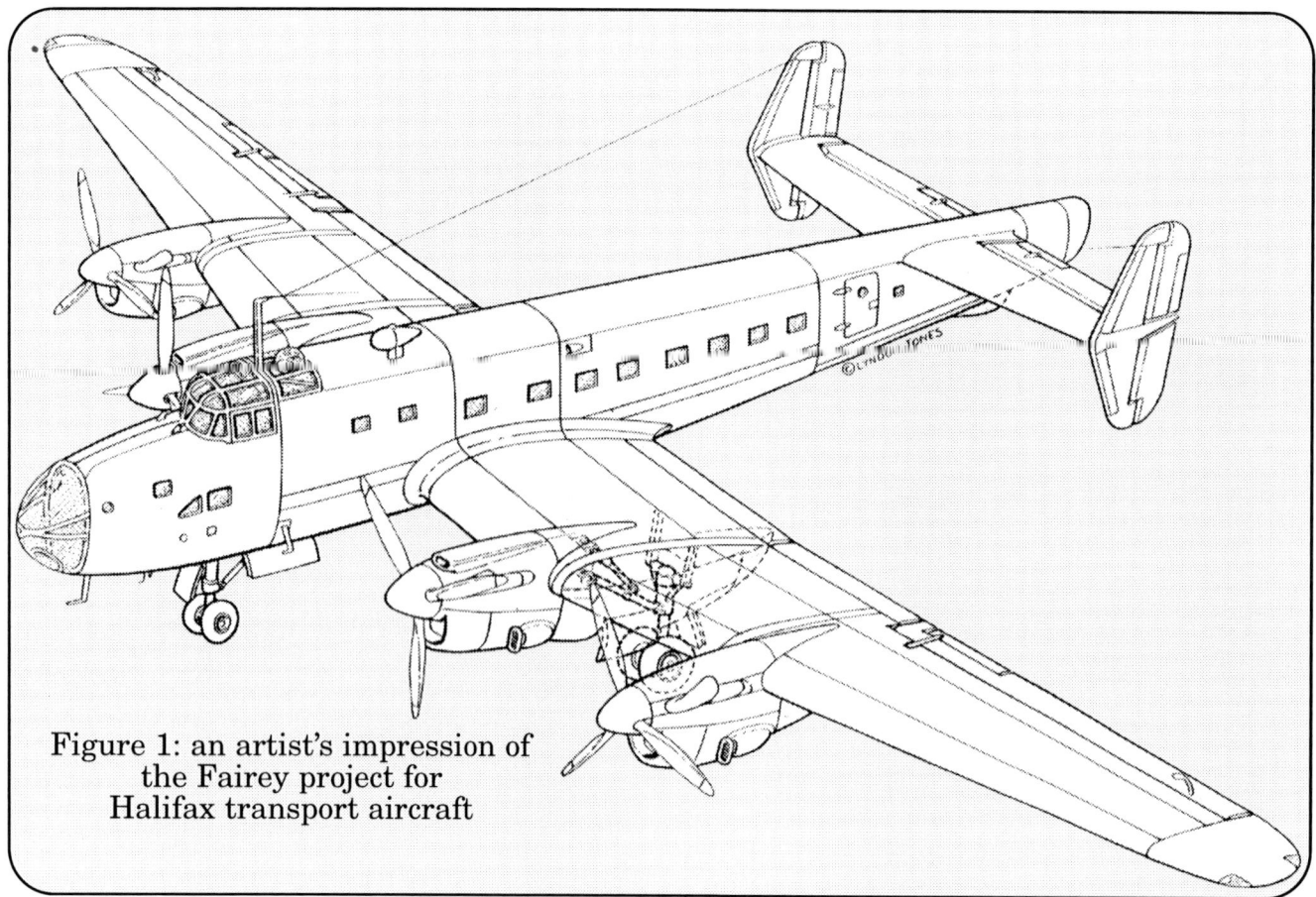

Figure 1: an artist's impression of the Fairey project for Halifax transport aircraft

transport project of the same period, which was rejected, the Avro project was given the go ahead and four prototypes ordered. This project became the York, the first of which had its first flight on 5 July 1942 at Ringway. An order for two hundred was placed and approximately fifty were built during the war years.

While the DC3/Dakota was a superb short haul airliner or transport, Britain was by 1943 considering schemes for postwar British designed and manufactured transport aircraft for long range use, and the DC4 would be the yardstick by which the contenders would be judged. At the same time as this need was being considered, a further U.S aircraft, the Lockheed Constellation, had taken off on its maiden flight on 9 January 1943 and was quickly conscripted as another military transport.

With the inauguration of the Brabazon Committee, papers and projects were sought from British manufacturers to cover a range of air transport for the post-war civil aviation world. In some of these papers on proposed designs the DC4 was used as a comparison. Some of the designs that would become hardware were the Avro Tudor, Airspeed Ambassador, and Vickers Viking.

While the DC4 and Constellation were designed as civil transports from their project stage, the Tudor, York and Viking were transport adaptations using bomber aircraft components, and thus penalised by the component weight and

Table 2 Comparison of basic data for three air transports.

	Douglas DC4	**Douglas DC6**	**Argonaut**
Wingspan	117.5 feet	117 feet	117.5 feet
Wing area, sq feet	1460	1463	1457
Length	93 feet 7	100 feet	93 feet 7 1/2 inches
Empty weight, lbs	43,400	53,623	46,832
Max loaded weight, lbs	73,000	93,200	82,000
Passengers	44	68	55
Maximum speed,	265 mph	301 mph	325 mph
Cruising speed (WM)	203 mph	269 mph	289 mph
Service ceiling, feet	19,000	29,000	29,500
Range, miles	1760	2810	3880

AIR TRANSPORT DEVELOPMENT

configuration. As originally proposed, the Tudor would have been (like the Handley Page HP64) just the replacement of the bomber fuselage with a circular section one of greater capacity.

The RDAP/MAP on 6 June 1944 forwarded to the 2nd Brabazon Committee a comparison of the Brabazon 3B (Tudor 1) and the DC4, that indicated that the Tudor 1 had a payload of 6263 lbs against the DC4's of 11,635 lbs. The Tudor being superior to the DC4 for the following reasons:

(a) higher cruising speed at the same percentage of take-off power (230 as against 203 mph).
(b) higher operating height (15,000 as against 10,000 feet).
(c) pressurised cabin for comfort.
(d) higher power loading giving 30% better rate of climb at sea level.

The contract for two Tudor I aircraft was placed in September 1944, followed by a production contract for fourteen Tudor I's on 1 November, which was increased to twenty aircraft in April 1945. The Tudor originated with a proposal by Avro on 18 October 1943, for a civil development of their Lincoln bomber for use on the North Atlantic as an interim airliner. On this basis the Lincoln's fuselage was to be replaced with a circular section pressurised fuselage. During January 1944 Handley Page received a contract for two Halifax conversions (HP64), which were to be built in accordance with specification C15/43.

In the USA meanwhile Douglas were developing the DC4 airframe, upgrading the engine power and increasing the fuselage length and capacity, which became the C 74 and first flew in September 1945. After this was developed the DC6 pressurised airliner that flew on 29 June 1946 at an all up weight of 93,000 lbs, this would be eventually increased to 97,200 lbs (see Table 2). The DC6 was the airliner that would be the one chosen by most airlines when it came to re-equipment with new aircraft in the post war world. The best features of the DC4 and DC6 would be embodied by Canadair when they produced their version of the DC4 powered by Merlin 620 series engines. This was the Canadair DC4M or C4 and was ordered by Trans-Canada Airlines (TCA) and the British Overseas Airways Corporation (BOAC), and known in BOAC service as the Argonaut.

By 1944 the U.S State Department had already prepared plans for the dominance of the post war commercial airline business by the USA, and in the following years it often appeared as if the British nationalised carrier, BOAC, was trying to help. The Tudor I made its first flight on 14 June 1945 and immediately ran into aerodynamic problems. Before the Tudor II made its first flight BOAC had placed an order for five Lockheed Constellations, whether this was a safeguard against failure of the Tudor is open to discussion, for the Constellation had suffered a number of problems and was during mid summer 1946 grounded by the United States Civil Aeronautics Board for 30 days after a fatal crash to one belonging to Trans World Airlines (TWA).

During this period the Hermes was under development and the Tudor I and II being tested and developed for BOAC. Came August 1946 and the skids were put under the British manufacturers by BOAC ordering six Boeing Stratocruisers, it was then obvious that further orders would follow for Constellations and Stratocruisers, as no major airline was going to operate such small numbers of each type. This eventually came to pass, for after the Tudor I and II had been involved in a series of modifications to reduce drag and improve performance, BOAC demanded over 300 modifications, a good proportion of which were cosmetic and had no bearing on the aircraft's technicalities The same pattern would be repeated on the Hermes, as told in Chapter 7, and whilst super critical attention was paid to the Tudor and Hermes aircraft, BOAC were prepared to accept American aircraft 'off the peg'. The final rejection of the Tudor aircraft was made on 11 April 1947, when the chairman of BOAC wrote to the Minister concerned. A court of inquiry was convened to investigate the Tudor aircraft and its cancellation, and although Avro received criticism, BOAC were castigated. Not that it mattered a great deal, for BOAC were granted government approval to purchase more American aircraft, the gate had been eased open and it was never again closed.

No one, including the author, would state with any sincerity that either the Tudor or the Hermes were outstanding civil transport aircraft, except in their structural strength, but fostered by the national carrier they could have been the stepping stones for more modern designs, for they had been classed by the Ministry as interim airliners. Even with American airliners BOAC were never in profit and always required a large subsidy. The airliner trend was in any case now to larger aircraft, pressurisation, tricycle undercarriages and modern low-drag aerofoils, and Handley Page with their Hermes IV design incorporated a tricycle undercarriage, and the latest in electrics and pressurisation. However, at De Havilland a new concept was taking place, this was their Brabazon type 4 with swept back wings and gas turbine engines. An advanced design, that once produced would consign the present American as well as the British airliners to antiquity, and would equip the RAF's first jet transport squadron (the world's first) - but that is another story, yet that jet aircraft in the fast transport role would replace the Hastings and Hermes - which are the subjects of this book.

The Hastings/Hermes conception.

The design of new aircraft should represent a large enough advance in technology and performance over present aircraft to prevent obsolescence occurring before it enters service, and the building of a prototype should be accomplished quickly so that modifications can be incorporated prior to production. Unfortunately, the Hastings, Hermes and Tudor could not claim to satisfy all of these requirements.

At this period, serviceability and reliability were not given priority consideration during the drawing up of the specifications, possibly because of the war and its demands for prominence in the performance and structural engineering fields. Yet surprisingly enough, a number of British aircraft and engine manufacturers were giving consideration to these aspects, and some would occur in the Hastings design.

From the project stage of the HP56 bomber to specification P13/36 until the production of the HP61 Halifax III, approaches were made or ideas formulated into the conversion of the design into a transport aircraft. This commenced first in 1937 when R.Verney of the Department of Technical Development (DTD) enquired from Handley Page the possibility of producing a transport version of the projected Vulture powered HP56. Handley Page rejected this, as it was considered that the structural weight and engine fuel consumption would make a transport version too heavy and uneconomical.

By the start of 1942 'HP' himself was considering the idea of a transport version of the Halifax, utilising standard Halifax mainplanes, tail unit and undercarriage in conjunction with a circular section fuselage. This culminated on 22 May 1942 with a memo' from 'HP' to G.R.Volkert suggesting that future policy for development of Halifax design work should consider a number of proposals. These mainly concerned the substitution of a large circular section fuselage, to have a diameter of at least 9 feet and incorporate a parallel portion of 12 foot length. Such as aircraft could be built in the following versions:

a) freighter for spares or bulky ground stores
b) aerial tanker
c) troop carrier
d) unpressurised short range bomber
e) pressurised high altitude bomber

With regard to the freighter version it was proposed that a large loading hatch roof was incorporated for bulky freight, such as Halifax spares which were in short supply at units! These proposals were discussed on 17 August with Air Marshal F.J.Linnell (CRD), but were rejected. Avro had by then been given the go ahead for their Lancaster transport, the York, with four prototypes being ordered, two with Merlins and two with Hercules. This was later changed to only one with Hercules.

Handley Page and 'HP' were not inclined to sit down and do nothing after Avro were given the go ahead for a transport version of the Lancaster, so on 11 January 1943 a proposal for a civil transport was submitted to MAP, but no encouragement was forthcoming from that area either. Meanwhile the Avro York had flown and Avro were getting upset by the number of requested modifications from various official departments, something that Handley Page had had to accept with the Halifax bomber when it was wanted for various roles. As well as sitting in on conferences on the York, BOAC had requested the allocation of five aircraft from the first production batch.

All this did not dissuade Handley Page from beginning preliminary design work under the designation HP64 and in May 1943 wind tunnel tests were in progress on the project. These determined that an increase in fuselage diametre from 9 feet to 11 feet would have little effect on stressing or overall drag. Preference for engines was for the Hercules as opposed to the Merlin, whilst Avro's was to the reverse. As Handley Pages were restricted by the amount of design work that they could undertake because of their Halifax programme, the design and manufacture of a rear fuselage for static testing was farmed out to Flight Refuelling Ltd, who had both factory space and drawing office facilities

The HP64 project was opposed by MAP, possibly because of their blessing for the Avro York, but also because the Air Staff had been conducting an exercise, the object of which was

Table 3. BOAC submission to Brabazon Committee.

	Modified York	**HP 64**
Engines	4 x Merlin	4 x Hercules
Normal maximum weight	63,000 lbs	65,000 lbs
Probable equipped weight	40,000 lbs	38,500 lbs
Maximum number of passengers	34	34
Weight of passengers/baggage	8000 lbs	8000 lbs
Cruising speed at 5,000 feet	190/200 mph	190/195 mph
Span	102 feet	104 feet
Length	86 feet 3 inch	81 feet 6 inch
Load at ultimate still range of		
1000 miles	15,150 lbs	16,250 lbs
1500 miles	11,650 lbs	12,950 lbs
2000 miles	8,430 lbs	9,850 lbs

AIR TRANSPORT DEVELOPMENT

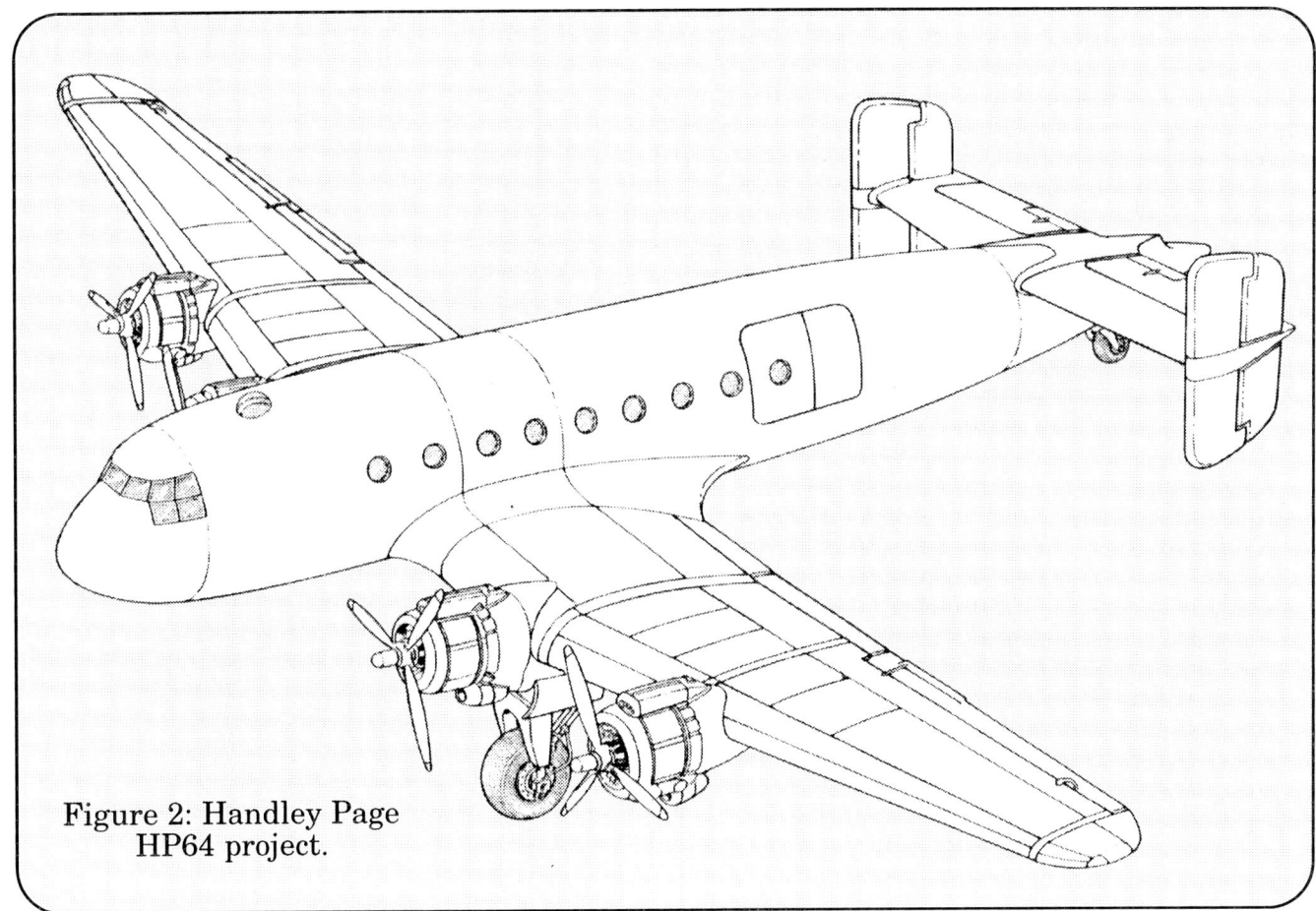

Figure 2: Handley Page HP64 project.

the use post war of bomber aircraft for trooping purposes. This resulted in projects 'Transport A' (a stripped bomber but still with turrets) and 'Transport B' (an unarmed transport version of a bomber). Nevertheless, the HP64 project (figure 2) had by then reached the stage where an approach was made to Roxbee Cox and Walter Tye of the ARB, to discuss civil airworthiness requirements that would affect the project.

In the summer of 1943 Handley Page were again having discussions with the Air Staff over the HP64 project, with a view to its use as a possible post war airliner and as a replacement for 'Transport B' in the troop carrier and military transport role. It would appear that the Air Staff were more interested in their two transport projects being produced than in future troop carriers, so Handley Page produced their Halifax transport versions for the RAF.

The Brabazon Committee had been formed to crystal gaze into the post war British aviation scene and in so doing they had formulated a number of 'Brabazon types' and set the basis for a successful return by Britain to civil aviation, that this did not come about completely cannot be blamed on the Brabazon Committee, but apportioned to a number of officials in various establishments and BOAC who dithered, as will be apparent as the text unfolds.

Numerous papers were submitted to the Brabazon Committee, both by operators and manufacturers, as well as by experts in various fields of aviation. One of these papers, No.44, was submitted by BOAC and gave a comparison between the York and the Halifax conversion (HP64).

BOAC in this paper pointed out that the Halifax conversion was a low wing monoplane and that the York with its high wing was more convenient for loading. BOAC's preference was obviously for the 'modified' York, for they terminated their paper with the point that the York was similar to the standard York, that was apparently to be the standard RAF transport aircraft and thus available in large numbers, whilst the Halifax conversion had no apparent advantages, but they did not mention that their proposed York was an enlarged one and not in production, and did not in the end go into production.

In November 1943 Volkert discussed his HP64 transport and HP66 bomber designs with both the DTD and DOR, but no decision was forthcoming. This was then followed by a visit to Cricklewood by DTD and DOR, when the production and designs at Handley Page were discussed in more detail. After which Volkert announced the priority list of new aircraft projects, these were:-

a) HP66 bomber
b) HP64 civil passenger transport
c) HP64 civil freighter version
d) HP64 military transport
e) HP70 Halifax C8

BOAC had in the meantime approached Roy Chadwick of Avro with a view to the possible production of the modified York with a longer fuselage. This came to the attention of N.E. Rowe at the Ministry, who totally objected to this proposal as Avro were already overloaded in producing the Brabazon Type 3B (Tudor) and York, without being committed to the design and production of an enlarged York. It was estimated at that time that the Tudor I would be flying in 12 months, but in actual fact the design work did not begin until June 1944 and September 1944 before a contract for two prototypes to specification 29/43 was placed.

With the issue of contract 3205/C4c to Handley Page for two Halifax conversions (HP64) for the carriage of passengers, 'HP' was enquiring over a possible version of it as an RAF transport aircraft, and was assured by N. E. Rowe that it was the broad policy requirements to have a transport version of the Halifax for use in the RAF sometime in the future, possibly thinking along the lines of the HP70 Halifax C8!

The next problem was to get an airline to sponsor the civil HP64, as the mock up had reached an advanced stage, so BOAC was approached by the Brabazon Committee to fulfill this function, although this airline had shown no interest in the project. After a number of weeks without a reply, Lord Brabazon specifically asked BOAC's representative on the Committee to get BOAC to make up their mind. This resulted with BOAC eventually being prepared to give Handley Page guidance as regards furnishings, passenger layout etc.

At a meeting at MAP the Minister concerned raised with W.Tye of ARB the question of weight limitations, hoping that they would not find it necessary to limit the weight to such a low figure that the aircraft would be uneconomical. Also made known at this meeting was the results of strength tests conducted at the Royal Aeronautical Establishment (RAE) Farnborough on examples of Halifax and Lancaster undercarriages, this was so that ARB could establish a standard for civilian aircraft undercarriages. The ARB were already concerned that the stressing conditions arising from a descent velocity of 12 feet per second gave low reserve factors in a number of points in airframe structures.

As can be seen, the change over from military to civilian aircraft construction was raising new tests and problems, as had been the change over from biplane to monoplane, and with one major national carrier there would be little opportunity to find other opinions on operating. A letter dated 10 December 1943 from BOAC to B. Shenstone of MAP gave some information on the discussions that were taking place between the Corporation and Handley Page. This briefly concerned a preliminary layout drawn up by Handley Page after discussions with the Corporation's Development Department and an alternative layout based on discussions between the Traffic and Development Departments of the Corporation. It also mentioned that Handley Page were investigating both proposals from a structural point of view as well as the sizes of baggage holds and doors. What was significant, was that the author of the letter, gave as his opinion that structurally the passenger and cargo versions would have to be different aircraft, it was a few years before airlines got used to hauling both freight and passengers at the same time.

At last on 1 January 1944 a meeting took place at Thames House London at which draft specification C15/43 was circulated, this referred to the civilian version of the HP64. A still air range of 2,000 miles was required, and the aircraft including its engine installations was to be suitable for operation in all climatic conditions. It was to be developed from the Halifax bomber by the introduction of a new fuselage and centre section. The passenger entrance was to be not less than 5.5 feet in height and 2.5 feet wide. The basic data supplied was:-

Wingspan	104 ft.
Wing area	1275 square ft.
Length	81 ft. 8 ins.
Height,	tail down 22 ft. 3 ins.
External fuselage diameter	11 ft.
Total load capacity	9,500 lb. passengers
	14,000 lb. freight
Number of passengers	34

On 14 March BOAC representatives visited Handley Page to inspect the mock up and design, and it was considered that the stage had been reached where the layout of equipment had to be decided on. This resulted in relocation of some equipment, but decisions were required as to the

Table 4 Basic comparison of three Handley Page designs.

	HP64	HP67	HP74
Wingspan	104 ft.	113 ft.	113 ft.
Wing area, sq. ft.	1275	1408	1408
Length,	81 ft. 8 ins.	82 ft. 1 ins.	95 ft. 6 ins.
Engines, Hercules.	Mk.100	Mk.101	Mk. 736
Empty weight, lbs.	38,500	48,427	55,350
Maximum all up weight, lbs.	65,000	80,000	86,000

provision of de-icing of the mainplanes, controls, propellers and carburettors. Some of the relocated equipment was the hydraulic panel, electrical panel and the environmental equipment. On 31 March BOAC wrote to the MAP requesting confirmation that a pressure refuelling system would be incorporated, this had not been called for in the original draft specification.

Also during March, Major Nicolls of Handley Page asked the Controller General of Civil Aviation that the HP64 aircraft be named the 'Hendon', but this was rejected in favour of the name 'Hermes'. The programme at Handley Pages called for the first HP64 to specification C15/43 to be built as a passenger aircraft, flying initially as an unfurnished shell and then fully equipped to airline standards. The second prototype to be built as a civil freighter with long loading doors on the port side, whilst the third prototype was to be identical to the second prototype but plus a paratroop door on the starboard side and provision externally for the carriage of twelve containers.

On 28 April 1944 Volkert urged 'HP' to settle future production policy in favour of either the HP66 bomber or the HP64 transport, as the manufacture of both could not be satisfactorily carried out at Cricklewood. He also felt that the production of a new transport project (HP67), combining the HP64 fuselage with the HP66 wing, would be preferable to either the HP64 or HP66. In conjunction with the HP67 he further proposed that the project should have a single fin and rudder to reduce interference drag from the large diameter fuselage.

In April the Air Ministry had issued specification C3/44 for a multi-purpose RAF transport, which would combine the roles of the Halifax C8 and A9. The HP67 project so closely matched the specification that it was tendered. Shortly afterwards the two HP66 prototypes were cancelled and two civil transport prototypes were ordered against specification C15/43 (which defined the HP64), with amendments to cover the change in mainplanes and tail unit. This aircraft project then became the HP68 Hermes, with the Hermes IA being a civil freighter version powered by four 1675 h.p. Hercules 101 engines, and the Hermes IB primarily a passenger aircraft powered by four 2000 h.p. Hercules 737 engines. On 31 July 1944 the CGCA confirmed that the HP68 would be known as the Hermes, that the first prototype would have a wingspan of 104 feet, but that the production aircraft would have a new wing of 113 feet wingspan, similar to the HP67 Hastings. In the end the first prototype also had a 113 feet wingspan.

An internal memo of 1 August in the Air Ministry spelt out a programme based on Handley Page planning, this mentioned the HP64 civil aircraft. It was expected that this aircraft would start its flight trials in December 1944, and though designed as a passenger aircraft would be despatched to the A&AEE as a flying shell with 104 feet wingspan. In regards to the HP67 military transport, Handley Page had been given a contract for jigging and tooling up, so that they would be able to swing into production as soon as the German War was over. Unfortunately, material could not be ordered until a definite prototype contract was placed, so that negated any contract for tooling up. It was also considered that the HP military aircraft airframe could be convertible into a civil passenger aircraft in any stage of its construction, as both would use the same airframe and same type of engine. The contract for the HP68 prototypes provided for the second civil prototype airframe to be taken from the HP67 production line, this appeared to be a large assumption seeing as no contract or production line for HP67s had yet been confirmed.

The decision to build two prototype Hermes Is at Park Street was taken in February 1945, when it was decided to complete the first one (G-AGSS, Drawing 3) as a flying shell and the second one (G-AGUB) as a fully equipped and furnished airliner. As there was never a surplus of draughtsmen at Cricklewood during the high point of the Halifax design, there had never been more than 90 employed to cover everything - the design of the single fin tail unit was sub-contracted to Blackburn Aircraft at Brough.

The latter half of the 1940s was still a period of learning, still problems with basic aerodynamics, as witness the problem of wing fillets on the Tudor and Hermes and the turbulence and drag at the rear of the Tudor's inboard engine nacelles. More than the military, civil aviation expected more for less, in other words, an increase in speed with very little increase in power, and drag reduction was the answer. In Great Britain the RAE at Farnborough were grappling with this and many other problems and researches, not just aerodynamics but airframes and engines. Some of it affected Handley Page aircraft, and RAE expertise was used willingly and thoroughly as acknowledged in Handley Page documentation.

This was not more so than in the change over from the twin tail to the single fin and rudder tail unit of the Hastings/Hermes. In regard to this an ex-Handley Page aerodynamicist did say that he felt in hindsight that they may not one hundred percent have interpreted the RAE report correctly, but that on the other hand there was some ambiguity in the way it was written. The single fin and rudder however turned out a success, but the tailplane and elevator were the cause of much research and experimental installation at Radlett, as related in Chapter 2.

Chapter Two
HP 67 Hastings Development

Initial Design.
On 22 May 1944 draft specification C3/44 was issued to Handley Page for a military transport prototype, this was to use the fuselage and tail unit of the civil transport aircraft to specification C15/43 and the mainplane and undercarriage of those of the bomber HP66 to specification B27/43 - except where design changes were necessary to comply with the requirements of C3/44.

This specification required the aircraft for use as a long-range freight or passenger transport and paratrooping use. This last use was only aired at the 2 June Advisory Design Conference, which brought forth from Sir Frederick Handley Page quick confirmation that incorporation of the paratrooping role would delay the production date by two or three months. At the same time it was pointed out that it would be quite straight-forward to introduce provision for paratrooping at a later date, and that retrospective modification action could also be taken. So it was agreed that the first production would not be penalised and that provision for paratrooping would be introduced as a major modification at a later date. It was requested that Handley Page produce prototype shell drawings by early September, with one prototype flying in 1944, although with a 104 foot wingspan.

The specification Hastings Mk 1/Pl was issued to replace the previous specification C3/44 and laid down certain requirements, some of which were:-
Fitted with four Hercules 100 (low drag) powerplants separate to engines.
24 volt single pole electrical system with four 'P' type generators.
Provision for auxiliary power unit.
Contractor's trials to show that the aircraft had been subjected to the flight tests for bomber aircraft as in AP970.
Still air range of 3,000 miles at 20,000 feet at most economical cruising speed with payload of 7500 lbs.
Cruising speed shall not be less than 230 mph TAS at 20,000 feet at not more than 50 percent take-off power when operating at full load.

It was also laid down that aircraft subsequent to the second aircraft would be regarded as production aircraft

During September 1944 a party from Handley Page paid a visit to RAE Farnborough to discuss the single fin and rudder tail unit, as this was a major break-away from the HP64 design. Mr Thomas of the Aero Department recommended that Handley Page use the unshielded horn balance of approximately 10 percent of the area of the rudder behind the hinge, in conjunction with a radius nose and balance tab. Then at a later date the balance tab could be replaced with a spring tab, in this way overbalance could be avoided. Regarding the tailplane and elevator, Thomas was of the opinion that a similar argument could be applied to the elevator as for the rudder, again advising on the proportions of horn and balance tab. Where the elevator horn was liable to come in and out of the slipstream there was still the chance of oscillations being set up and of distortion of the horn due to localised loads. It was also agreed that the particulars of the ailerons would be sent to RAE for them to check.

The drawings of the HP67 had by then been completed and the decision made that the wingspan would be 113 feet using the HP66 wing, yet no decision had been conveyed to the company regarding the wing de-icing, so they decided to go ahead with building into the wing the necessary parts. By December eight Hercules engines had been delivered to the firm, four of them for trial installation in a Halifax. This was followed by the final mock-up conference on the flight deck layout.

The next big job was the manufacture of the specimen test wing, this also included thirty feet of the fuselage (frame 240 to 820) attached. A Works Order was issued and Boulton Paul undertook the production of drawings for the test wing and jig. McKenna of the drawing office suggested that MacRostie undertook the production of the test wing at Radlett, as Cricklewood only produced separate fuselage and wing sections. The wing structure was to be strengthened for clearance to 75,000 lb maximum weight. During July 1945 Handley Page members visited the Messier works [Post-war Messier became Electro-Hydraulics]. to discuss and decide on tests required for the Hastings' undercarriage, and Bristol Aero Engines notified the firm that they were re-designing the low-drag powerplant for improved cooling as well as modifying it for Hastings installation. Further work on this would of course help in the development of the Hermes powerplant, but did also contribute to a delay in delivery of the later marks of Hercules for the Hastings.

Handley Page's costing for the two Hastings prototypes gave a provisional figure of £350,000, the first ten aircraft at £22,000 per aircraft, and from the 21st aircraft £42,000 per aircraft. This for an aircraft with an initial maximum all-upweight of 75,000 lbs., rising to 80,000 lbs., was competitive. Post-war the British aviation industry had, amongst its other postulations, given a review of production costs. This was estimated for an aircraft of 60,000-70,000 lb. loaded weight that

HP 67 HASTINGS DEVELOPMENT

the selling price would be £2.75 per lb. (£192,500 each), with the development costs being approximately £31 per lb. (£2,170,000). With regard to the Hermes the Government contributed £1.3 million to the costs of it from project to hardware, which must have helped in the competitive prices of the Hastings and Hermes, being of similar construction. Yet, before the decade was out, up to ten times that amount was contributed to other firms for aircraft projects.

In early 1945 the firm proposed a lengthened version of the Hastings known as the HP73, this would have had a fuselage length similar to the Hermes II. Discussions on it dragged on for months without any decisions being made, even though the original suggestion had crept out of the Ministry; until eventually in March 1946 the DOR decided against ordering it.

By March 1946 the prototype Hastings, TE580, was being prepared for its final inspection by AID, the Hermes I having crashed on its first take-off. TE580 was stripped down into main sections and transported by road to RAF Wittering, where its final assembly was carried out prior to its first flight. This was made on 7 May by Squadron Leader Hartford, taking off at 1240 hours a flight of 35 minutes was carried out. During this flight the aircraft was taken up to 11,000 feet and cruised at 200 mph. The effect of flaps and undercarriage was tried on handling and on elevator control, rudder control was also checked with outer engines throttled back in turn, all were satisfactory.

Even during this testing period the choice of a single fin and rudder and its stressing was causing both correspondence and wind tunnel tests. Cedric Vernon, Chief Aerodynamicist at Handley Page, in a letter to the Director of RAE Farnborough, quoted the results on a Hastings in a wind tunnel. The tests were to determine the torque on the fuselage due to asymmetric distribution of tailplane load that occurred in yawed flight (the same loading which were assumed to have led to the Typhoon accidents). Tests were carried out with the tailplane at zero, positive and negative dihedral. From the tests and arguments submitted there appeared to be differences between Vernon's results and the implications of the relevant part of AP970 and the RAE Typhoon report. So Vernon requested from RAE confirmation of some of their report's statements, the results of which were open to interpretation.

The second prototype was TE583, and this carried out its first flight on 30 December 1946, again no handling difficulties were experienced, the fin and rudder being satisfactory. Although no full testing to the aircraft's limits were indulged in, the results at that period appeared to confirm satisfactory co-operation between Handley Page and RAE Farnborough. The first production aircraft, TG499, carried out its maiden flight on 25 April 1947 flown by Hazelden. Almost immediately tail vibration was experienced, that was quickly cured by 'cording' the rudder trailing edge, to be followed later with modification 321. The vibration was traced to spring tab flutter at high speed.

The following month saw TE580 take to the air on the 22nd, with its engine's thrust lines tilted up 2.5 degrees, as a means of overcoming the type's longitudinal instability at 78,000lbs. all-up-weight with CG aft, unfortunately this was not the final solution. What is surprising is that within twelve months of the type's first flight the aircraft was being flown at 78,000lbs., 3,000lbs. over its proposed weight limit.

Bristols were by then discussing improvements to the powerplants, as well as sorting out problems that had arisen during the early test flights. From these were introduced an improved ignition harness which was more resistant to moisture, a new plug called 'Long

Handley Page Hastings prototype TE580 on one of its early flights. [E Lyons]

HANDLEY PAGE HASTINGS AND HERMES

Frontal view of prototype TE580 with all turning and churning! [J. Knivett]

Tom', alterations to the rear exhaust manifold and improved cylinder cooling for production aircraft. It was also established that early high cylinder head temperatures were due to faulty thermocouples. In any case, as the penalty for opening the cooling gills was small, it was reasonable to rely on the cooling gills to keep the cylinder temperatures within the limits during the climb, but recommended that the CHTs were watched during the trials at Boscombe Down.

By 20 August 1947, 95 percent of the total quantity of jigs and tools for Hastings' production had been completed, this amounted to a total number approaching 15,000, and it was estimated that the total tooling cost at that time would amount to £735,000. Already enquiries were arriving at Handley Page's sales office about a civil version, AVM. D.C.Bennett of British South American Airways on 23 September spoke direct to 'HP' in regard to the purchase of civil Hastings for the company. By the end of the year QANTAS were making enquiries through BOAC, and in January 1948 Handley Page received a request from BOAC for them to carry out handling tests of a Hastings in order to render a report to QANTAS. In April, QANTAS advised the Australian Air Ministry of their desire to operate Hastings, subject to Ministry tests. The actual flight assessment was carried out by Captain Alderson of BOAC with 'Sandy' Sanders of Handley Page covering him.

The flying was carried out on 4 and 6 February at Hatfield airfield and tests were quite extensive, including climbs on three engines, baulked landings and flight in cloud with bad turbulence. Captain Alderson's comments indicated that he liked the aircraft, it was a straightforward machine, and easy to handle, and he was impressed with the cockpit layout. His adverse comments covered the entrance door in the floor, the noise in the cockpit was greater than he had anticipated, while the ailerons appeared to be a little too inneffective for comfortable control with one hand during approaches in cross-winds.

Tests were still proceeding at Radlett, and in October 1947 Stafford had established through development work that the bulbous-nosed elevator as well as the flap to elevator tab interconnection were essential improvements, irrespective of tailplane form, and were to be incorporated on new production aircraft. Already in production aircraft as well as TG580 were being flown with different tailplanes to overcome A&AEE's criticism of stick-free longitudinal instability and elevator heaviness.

At RAE Farnborough the Hastings fuselage with stub wings had undergone tests to destruction on the 'Cathedral'test-rig, and full down load had recorded 5 percent over the design load. This pleased 'HP' so much that he wanted a coachload of supervisors to see the test specimen and its strength. The main failure was between the underneath hatch and the rear spar, with the remainder of the fuselage in remarkably reasonable condition. With the full factored load on the fuselage the main doors could still be opened and closed. 'HP' very sensibly suggested that the fuselage be repaired and further testing carried out on it to determine any other weaknesses, a procedure which would become standard practice.

With the production order for the Hastings

Table 5. Basic performance data of Hastings civil freighter.

Weights, basic equipped	49,000 lbs
maximum take-off	78,000 lbs
maximum landing	70,000 lbs
Wing loading at take-off eight	55.3 lbs/sq.ft.
Power loading at take-off weight	11.6 hp/lb.
Speed, max WM at sea level	241 mph
max WM at 23,300 ft	298 mph
max WM at 25,000 ft	284 mph
Rate of climb at sea level	1025 ft per minute
17,000 ft	773 ft per minute
25,000 ft	175 ft per minute
Time to 10,000 ft	9.8 minutes
Time to 20,000 ft	20.7 minutes
Service Ceiling	26,000 feet
Maximum cruising speed in WM at 10,000 feet	276 mph
Maximum range speed (2,000 miles)	212 mph
Maximum range	3,159 miles

C1 established and a Mk.C2 envisaged, incorporating all the improvements developed through static and flight test work, it was hardly surprising that a further version was projected, especially in view of the contact between the firm and Service personnel at AFEE and A&AEE. The newest project was the Hastings VI with rear door-cum-loading ramp and powered with Hercules 763 engines. A meeting was held to discuss certain features with the object of obtaining a reaction from the R&D Airborne personnel; these features included a longer fuselage, tricycle undercarriage with fourwheel bogies on the main undercarriage, and airborne support equipment. Apart from establishing that the main feature of a rear loading ramp was essential and an improvement on the Hastings side door, little else was established. Handley Page staff considered that

TR580 on a test flight with the two port propellers feathered to test handling [J. Knivett]

The crew and respresentatives in front of TG 503, just before departure to Australia and New Zealand [J. Knivett]

the Army representatives appeared to show little interest and regarded this Hastings development as an interim aircraft. Further to this, Handley Page could not ascertain whether any aircraft specification had been issued covering Army Airborne requirements. The project was designated the HP89 with the data in Table 6.

It was established that the basic structure of the Hastings VI would be identical to the Hermes IV, except for the rear doors, ramp and paratrooping door. As it was, nothing came of either the civil Hastings freighter nor the Hastings VI, but interest was being shown by the RAF in the Hastings C2 and C4 VIP aircraft. For, inspite of the CG limitations of the Mk.Cl, the RAF were finding the Hastings a useful aircraft of high transitting speed. The biggest problem with it, as compared with tricycle undercarriaged aircraft, was the sloping cargo floor.

TG503 now comes into the picture, for as part of the A& AEE test programme it was to carry out a route and extensive flight test. This commenced on 11 March 1948 on a flight to Australia and New Zealand. The tour lasted four

Arrival of TG503 at Paraparaumu, New Zealand after the flight from Britain [K. Lyons]

HP 67 HASTINGS DEVELOPMENT

months and the aircraft was crewed by RAF and Handley Page personnel, who were accompanied by MoS observers. The route was from RAF Lyneham to Sydney, Australia, and the flight took 46.5 flying hours, with an average speed for the whole flight of over 250 mph, at an average take-off weight of 77,600 lbs. The significance of this tour was that it was the first time that any new RAF aircraft had undertaken so extensive a tour prior to entering service.

The outbound flight terminated at New Zealand, the Hastings landing at Paraparaumu airfield, about 30 miles from Wellington. It was thought in New Zealand that Paraparaumu would be too small a size, as well as the surrounding hills making the approach difficult, but the Hastings had no difficulty landing there. It also gave several demonstration flights there, that illustrated its ability to land in 1,000 yards or so after a 1 in 3 gradient approach at a weight of 67,000 lbs.

On the return flight the opportunity was taken of giving demonstration flights in a number of countries including India and Pakistan. The whole tour was satisfactory and supplied a great deal of useful information on the handling of the aircraft away from base, as well as its performance under tropical conditions. Faulted was the inadequacy of fuel off-loading, the fuel filler neck was too small for fuelling, as well as the cabin cooling being inadequate for tropical conditions.

Hastings TG503 flying past the Clock Tower at Hastings, New Zealand. [K. Lyons]

As a result of the flight to New Zealand, the country's government ordered four Hastings C3 aircraft for the RNZAF. The C3s were serial numbered NZ5801 to 5804 inclusive, and were powered with Hercules 737 engines, which gave a quicker take-off and faster speeds at low altitude than RAF Hastings at those altitudes. They were also equipped with additional modern radio, the RAF Hastings at that period still using the rather dated TR1154/1155.

During normal production testing of TG537 on 5 November 1948 the opportunity was taken to carry out five tests to check the stalling characteristics with all propellers feathered. Flight Lieutenant Broomfield was the test pilot and his report on the five tests gave the following:-

(1) Stall clean [with the undercarriage and flaps retracted] with outer engines feathered, inner engines throttled back, stalling speed 100 knots.

(2) Stall clean with all engines throttled, stalling speed 101 knots.

(3) Stall clean with all engines switched off and propellers windmilling, stalling speed 103 knots.

(4) Stall clean with two inner engines feathered and outer engines throttled, stalling speed 106 knots.

(5) Stall clean with all propellers feathered, stalling speed 109 knots.

The stalling characteristics were the same in all cases. The control column was pulled back to the rear stop and held there, the nose of the aircraft continued to pitch with increasing violence and buffeting of the elevator grew severe, recovery was normal and immediate. Stalling the aircraft clean with all engines stopped and the propellers feathered hardly indicated an aircraft with serious problems. The Hastings was accepted by the MoS against the recommendations of the A&AEE, who required further alterations to the aircraft before acceptance. So was the A&AEE setting too high a standard as regards controls, as witness the accident to TG574 on 20 December 1950, when it was flown without rudder and elevator after the controls had been severed.

Even after acceptance into the Service in 1949 within the limits imposed by A&AEE in their normal manner, testing at Radlett continued to attempt to rectify the deficiency in longitudinal stability determined by A&AEE. Commencing in the summer of 1947, TE580 had been the test vehicle for numerous experiments with the tailplane. These tests continued through the winter, when during this period the tailplane was set at 5 and 10 degrees dihedral, as well as 5, 10 and 15 degress anhedral. All these tests being witnessed by a camera mounted on the top third of the fin, which photographed the flow behaviour of wool tufts fitted to the tailplane leading edge. None of this had the desired affect and so the idea of incorporating anhedral or dihedral was dropped.

Table 6 HP89 Hastings Mk VI basic data.

	Load in lbs	range in miles	fuel in galls.
freighter	19,085	1280	1272
vehicle dropping	19,085	1280	1272
supply dropping	17,917	1250	1434
ambulance	18,865	1315	1303
paratrooping	19,038	1030	1279
troop transport	18,232	1420	1391
exterior pannier	15,792	1640	1730

Weights,		
	total structure	22,615 lbs
	four Hercules 763 engines	16,565 lbs
	tare	46,870 lbs
	basic operational equipment	50,655 lbs
	landing	73,000 lbs
	take-off	80,000 lbs

In May 1948 TG502 joined the test programme, having its tailplane bodily lowered 16 inches below the original position and its incidence increased by 2 degrees. This was part of the practical application of wind-tunnel work undertaken at Radlett under Stafford's control. As so often happens, practice does not always follow theory, and in this case although there was an improvement in longitudinal stability, this was accompanied by a deterioration in the powered approaches, and aggravated by large trim changes with power changes, plus an increase in stick forces on landing.

Next into the test programme in November was TG501, this had been fitted with an interconnection of flaps and trim tab, and a reduction of the pilot's elevator trim gearing. Testing revealed that although the change in trim with flap operation was substantially reduced, there was inadequate stick-free longitudinal stability at low speeds and high stick forces and change of trim with power on the landing approach. As regards rudder 'tramping' and lateral shuddering, this was still apparent.

By February 1949 TG502 had again been modified, this time having an elevator and tailplane with increased span and area, the areas being increased ten percent and the tailplane lowered 16 inches. There was also a reduction of 2 degrees in the tailplane incidence, a new nose shape to the elevators, as well as a spring tab on the elevators. Testing revealed that the longitudinal control was then satisfactory with a major improvement in elevator control forces, but the ailerons were still too heavy and out of harmony with the other controls. By June TG502 was listed as a Mk.C2 and had been fitted with, amongst the other modifications, sharp entry root fairings (to be introduced on all later aircraft). When tested, the aileron control was found to be much improved, having become lighter and effective throughout the speed range and in harmony with the other controls, confirmed by

TE580 fitted with a tailplane that has 15 degrees dihedral [Handley Page Association]

HP 67 HASTINGS DEVELOPMENT

Above: Hasting Mk.i1 TG502 with wide span tailplane, in it's original high position.[JE. Watts].
Below: The prototype TE580 on test with widened tailplane and anhedral, still in the high position [E. Watts].

A&AEE. The rudder at low speeds was lighter and less effective and at higher speeds it was heavier. This was the control pattern set for the Mk.C2 onwards.

In regard to an order for the Mk.C2, by March 1949 the delay in placing a contract was affecting production at Handley Page, even though the firm was itself in a good position regarding the development of a large jet bomber. Sir Alec Coryton realising the dated position of RAF Transport Command, brought this contract position to the attention of the AMSO and asked that this matter be treated as urgent. The Air Ministry was as usual in the hands of the Treasury - and a Socialist Treasurer at that. The engines that Handley Page required for the C2 were the Hercules 260 series, which would confer on the C2 an all-round improved performance; unfortunately these engines according to Bristol were unlikely to be available until early 1950.

A liaison meeting on the 13 May was held to discuss the C2 and C2A VIP (later designated the C4). The decision was made to incorporate a common fuel system for both marks of aircraft and to utilise the same centre wing and spars as the Hastings C1, with the intermediate wing structure as for the C1, and all oil tanks of increased capacity. It was decided to change the outer wings using a front spar of DTD364 and rear spar of DTD683, both spars being machined to conform to the Hermes IV dimensions. The structure was to be similar to the Hermes IV between the spars,

TE580 on test with widened tail and five degree dihedral. Note the camera mounted on the fin. [E. Watts]

HANDLEY PAGE HASTINGS AND HERMES

TE580 with widened tailplane in original postition, no dihedral. The tailplane is tufted for airflow tests and there is a camera mouted on the fin [E. Watts]

with the tailbox similar to the C1 but with Hermes IV aileron hinge ribs. The ailerons were as for the C1 but with lightweight spring tabs and trim tabs and reduced mass balance.

Based on RAF requirements for Hastings aircraft up to March 1951, it was decided in May 1949 to increase the present order by 65 aircraft, but to be delivered to C2 standard. Following this on the 10 June a decision was made to call up a new specification (l9/49/P) for the C2. This called for, amongst other things, tailplane to be lowered 16 inches and of increased span, spring tabs for elevators, increased fuel capacity by approximately 600 gallons in flexible tanks in outer wings, and the introduction of a quartermaster's station. Within three months of the draft specification being issued, an amendment was issued, this called for Hercules 106 engines, protection against uric acid (a problem just recognised on transport aircraft), and an investigation into vibration.

TE583 on the 14 June 1949 returned to Radlett, as it had been allocated for the installation of Sapphire jet engines in the outer positions. It was on loan for six months, which was to include the testing time. After going to NGTE for engine testing and to Bitteswell for further work by Armstrong-Siddeley, it eventually arrived back at Radlett on 14 May 1952. It was then fitted with a Victor type escape door, going on to A&AEE Boscombe Down afterwards for further flight testing.

During previous discussions at the Air Ministry, it became patently obvious that the rather tired VIP Yorks were hardly the type of air transport that one could expect VIP members of the Forces or Government to pay official visits in. So the RAF put forward proposals that resulted in the issue of specification C.115/P, this called for a special version of the Hastings C2. The main requirements of this VIP version affected the furnishing, soundproofing, accommodation and facilities. This resulted in Handley Page producing the HP94, the Hastings C4 - all mod' con' that included the kitchen sink.

Further projects and studies on the Hastings' theme would be considered, these included prop-jets, lengthened fuselage with tricycle undercarriage, pressurised fuselage - but in general it was realised that the Hastings/Hermes formula was played out. The pure-jet and prop-jet transport with modern aerofoil sections was making their appearance. Yet as one Hastings' pilot put it "Even if the Hastings had done nothing else, it at least introduced a decent crew cabin and comfortable seats to Transport Command". It had contributed more than that, as Chapter 6 will show, it gave to the RAF long-legged air transportation with comfort and speed, which was only bettered with the introduction of the jet-transport.

TE583, the second prototype in 1949, having had A-S Sapphire engines fitted in the outer nacelles [E. Watts]

Chapter Three
Production

Production of the Hastings was carried out in the manner previously used in the manufacture of the Halifax, detail parts and sub-assemblies being produced at Cricklewood, with the final assembly being carried out at Radlett. It was the procedure for close contact to be maintained between the main design office, jig and tool design office and the factory floor supervision. In this manner the jig office was able to supervise the design of tooling and maintain direct contact with the factory foreman. So the chief jig and tool draughtsman was able to advise as well as to discuss with the appropriate design draughtsman the evolution of the design. In this way, with every design, the design was suitable for production from the start, and advanced information on it could be passed direct to the planning office.

At Handley Page the application of the split construction method began on the Harrow of 1935 and was further developed on the Hampden in 1936, Volkert having introduced this method as well as photo-lofting into Handley Page after a visit to the USA. As related in the author's *'Halifax - Second to None'*, all Handley Page aircraft designs commenced with an exploded view of the components, this and the split construction method was continued with the Hastings aircraft.

Handley Page policy was to use full scale lofting, and the transfer of the dimensions from templates, as well as the pre-drilling of components, such as frames and stringers, from templates and jigs. These items or components were then assembled or riveted together, before transfer to the final assembly jig for the riveting on of the skin, the final assembly jig being the only jig used on assembly. The greatest possible advantage was taken of these systems to use the prototype as a means of developing the basic tooling for the production aircraft. In this way, by the time that the prototype was completed, a complete system of tooling had been developed and duplicated, minor changes being introduced as necessary on the duplicate tooling.

Against this background both the Hastings and Hermes were developed, and as both types were in many respects very similar, there was much similarity in construction between the two types, and this meant that a large amount of tooling was common to both types of aircraft. For instance, the Hermes IV fuselage was 160 inches longer than that of the Hastings, whereas the Hastings had a tailwheel undercarriage and a much stronger floor, yet the basic structure of both types was similar.

Stringers and frames of both types were pre-drilled, so that when assembled on the fixtures they could be immediately riveted up. The floors of both types consisted of Plymax supported on floor beams positioned at the frames, the floor beams being supported by two bracing per beam on the Hermes and eight bracings per beam on the Hastings (see drawing 7). The Hermes IV, like the Hermes II, had a fuselage length extended by 160 inches above the Hermes I length, but the 160 inches was disposed differently on the Mk.IV, being 60 inches forward of the wing and 100 inches aft. This disposition was decided on to meet BOAC's requirements, that called for a certain weight distribution. The fuselage construction began with the floor structure that was built up from pre-drilled components. These were then placed on a horizontal fixture, with the longitudinals and beams laid out in an inverted position and with the intercostals service bolted

Construction of Handley Page Hastings fuselages on mobile fixtures [Handley Page Assn]

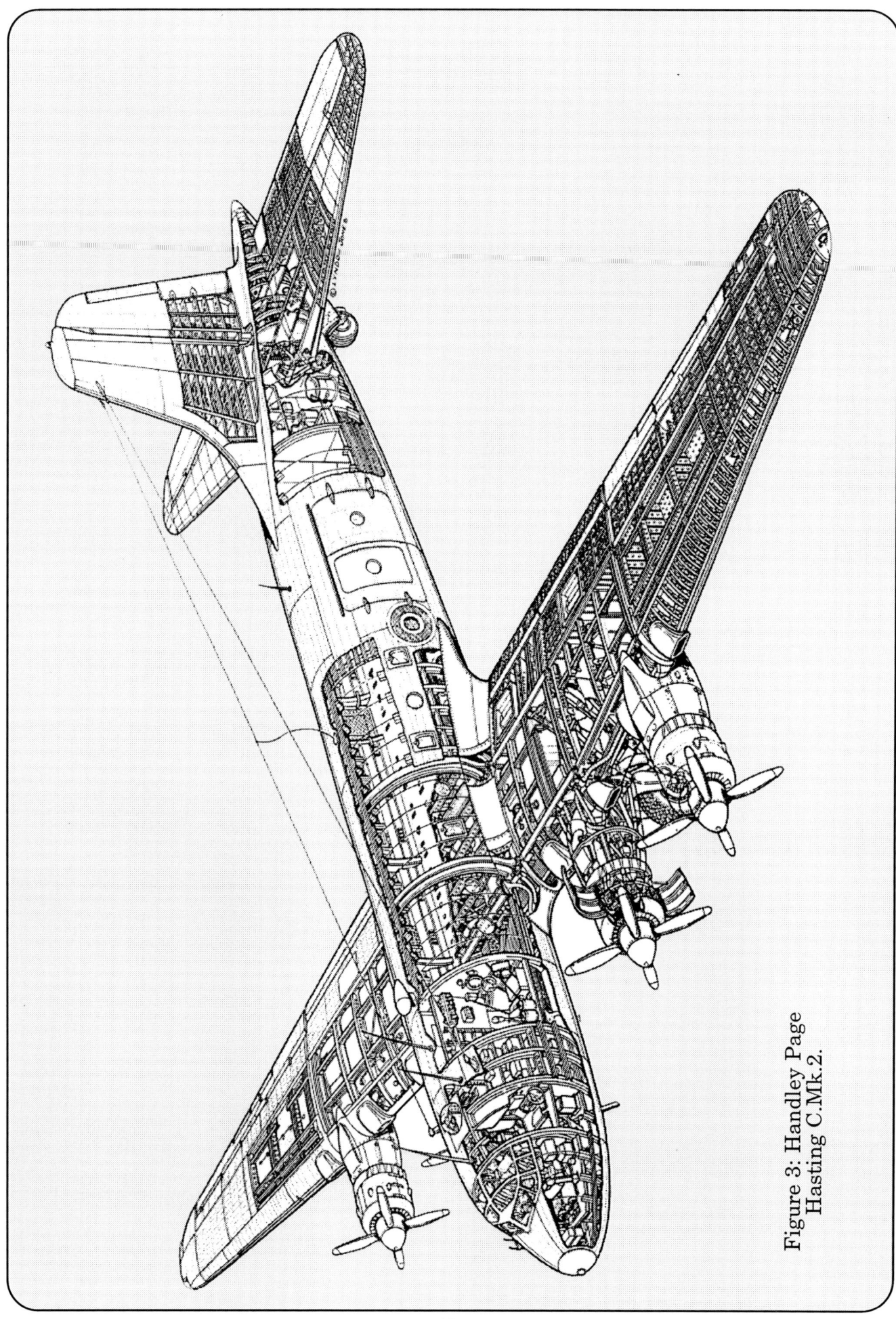

Figure 3: Handley Page Hasting C.Mk.2.

PRODUCTION

Hastings final assembly at Radlett, with a centre-section mainplane in the foreground. [Handley Page Assn]

together. The Plymax floor was then fixed and the whole formed a rigid basis on which the frames were erected.

The production of the sub-assemblies at Cricklewood was achieved by the assembly of as many sub-assemblies outside the jigs, each one as fully equipped as possible. The fuselage frames were received at Handley Page in the form of rings with the ends overlapping. These were first rolled to remove any distortion, then they were placed in a fixture to check the correct diameter and cut to the correct size, this being followed by being placed in a master jig and drilled. The horseshoe frames of the 'covered wagon' section of the fuselage/wing joint were of extreme importance, requiring extreme rigid construction and were involved with establishing the accurate assembly of fuselage and centre-section. These were assembled in a drilling jig where the flanges and webs were drilled simultaneously. The floor assembly was next transferred to a mobile fixture where it was riveted and bolted up, followed by the frames being attached. These mobile fixtures were used for the front and rear fuselage assemblies, the whole being constructed so that it appeared as a basketwork before the skin was attached. At this stage a floor to frame temporary bracing was attached to maintain the structure rigidness and correctness. The procedure for assembling the skins to the structure was fairly straightforward, the skin panels being received undrilled were drilled from the existing holes in the structure.

The two-spar centre-section contained and was part of the integral fuselage. This portion included the lower frame segments, floor beams and floor, with the lower portion also including

Table 7. Hastings and Hermes Production.

Aircraft type	Specification	Qty	Serial numbers
Hastings prototypes	C3/44	2	TE580 & TE583
Hastings C1	C3/44	39	TG499 to TG537
		37	TG551 to TG587
		24	TG601 to TG624
Hastings C2	C19/49P	25	WD475 to WD499
		17	WJ327 to WJ343
Hastings C4 VIP	C115/P	4	WD500, WJ324 to WJ326
Hastings C3 for New Zealand		4	NZ5801 to 5804
Hermes Mk I	C15/43	1	HP68/1 G-AGSS
Hermes Mk 2	C15/43	1	HP74/1 G-AGUB
Hermes Mk IV	37/46	25	HP81/1 to HP81/25
Hermes Mk V		2	HP82/1 and HP82/2

HANDLEY PAGE HASTINGS AND HERMES

part of the side skins, to be attached to the main fuselage skins. The fuselage centre-section and joint was not a transport joint, it was only for assembly, and once the fuselage was completed it could not be dismantled. The fuselage subsections were transported along with the wing sections etc to Radlett airfield for final assembly. Once the fuselage sub-assemblies were completed in their skin covering fixture, they were transferred onto a cradle trolley.

The mainplane structure of the Hermes, Hastings and Halifax were of similar basic construction, with the front spar consisting of massive channel section booms of DTD364 for the Hastings and DTD683 for the Hermes. Machined fittings bolted to the front faces of the booms form attachment for the tubular members of the engine mounting and for the undercarriage pivot points on the Hastings. On the Hermes IV the engines were positioned higher in relation to the wing datum than on the Hastings, and the undercarriage pivot points were mounted on large channel section machined fittings bolted to the rear spar. The leading edges and trailing edges of the mainplane were built up as separate sub-assemblies. The sub-assembly breakdown was carried on throughout the operation and was apparent in the construction of the mainplanes.

In the case of the assembly of the wing centre-section and intermediate wing, the front spar was set up in a fixture. The ribs were then assembled to the spar, followed by the tank bearers. Engine ribs were also installed at this stage and the rear spar bolted to the ribs. Top and bottom skins were received pre-drilled at the service bolting positions only and where-ever possible the spanwise stiffeners were pre-assembled to the skins. Drilling of the skin to the structure was then carried out in situ and the skins riveted on. The outer wing assembly differed a little from the previous as the leading edge was assembled to the front spar first, the whole was then set up in a fixture. The aileron hinge ribs were next assembled to the front spar and the rear spar placed in position, followed by the light ribs and tank ribs being next bolted into position. The whole assembly was then trued-up and the spanwise stringers attached. The structure was then transferred to the skinning structure where the rear spar straightness was checked, all bolts tightened and the top skin applied, followed by the lower skin.

The method of tooling, pre-drilling and fixtures and the reduction of the use of main assembly jigs was described in the 'Aircraft Production' magazine in 1949. In there, in referring to the construction of the Hermes and Hastings, the author made the point that it would be difficult to find other aircraft of comparable size so simple to produce in the assembly construction stages. This same method of construction was also continued into the production of the tail unit and control surfaces. The fin and tailplanes were assembled out of fixtures, being assembled in a vertical position without positive location except a vertical upright. One interesting point of the assembly system concerned the tail unit, for both the Hermes and the Hastings used most of the same pre-drilled items, so assembly of the two types was achieved by eliminating the items not specific to the type. Each type of unit was assembled to a fixture for skinning, spar straightness was checked and adjusted where necessary, then the skins were applied and riveted on. The remaining work, such as assembly of the controls and de-icing equipment, was carried out on floor assembly.

At Radlett was the final assembly and a presswork shop and foundry. The final assembly of the aircraft commenced with the fuselage being first assembled to the centre-section mainplane by means of bolted attachments at the horseshoe frames and spars. The nose section was next offered up and bolted to the curved channel section at the front spar. Splicing was then made of the fuselage stringers and the final skinning carried out. After this the intermediate wings, outer wings, tail control surfaces and wing control surfaces and engines were fitted, followed by the control runs being connected, equipment installed and the joining up of the services. This was followed by the final inspection, check of controls and services, engine runs, clearance by the AID and the aircraft was ready for its test flight.

HANDLEY PAGE HASTINGS AND HERMES
Chapter Four
Testing and Analysis

The testing programme of the HP67/68 was to commence with the flight of the Hermes Mk.I prototype G-AGSS, Taxying trials that commenced on 1 December 1945 caused the brakes to overheat, so the wheels were changed and the brake pressure reduced, The following day further taxying tests were carried out in the morning although the weather was bad, when Flight Lieutenant Talbot, the firm's test pilot, complained that he was experiencing great difficulty in getting the tail up during these runs. At no time during these taxying trials did the tailwheel leave the ground, except when the pilot throttled back and was braking hard. Then the weather deteriorated further and so the taxying trials were called off. Shortly afterwards there was a distinct improvement in the weather and Talbot and 'Ginger' Wright (flight test observer) were recalled to carry on with the tests, Talbot arranged for groundcrew to be at the far end of the runway in case he decided to offload some of the ballast, and Wright got two other flight test observers, Brailsford and Steel, to film the tests with a 16 mm camera.

No mention had been made of carrying out a flight and it was thought that there would more taxying and possibly a 'hop'. So that when the aircraft did take-off it came as a surprise to most observers, including R. S. Stafford, Handley Page designer. During the run Talbot accelerated the aircraft and it appeared to be thrown into the air at the hump in the intersection of the two runways. Observer's stories all appeared to agree in that the tail appeared to come up after some time, then went down with a thump. The wheels then left the ground and immediately the aircraft commenced a short period pitching oscillation, that gradually increased in amplitude with the aircraft climbing very slowly. Oscillation was then controlled and the aircraft continued to climb and turn very slowly to the left, immediately the oscillation commenced again. It then appear:ed as if the aircraft reached a vertical attitude, fell on its back and rolled into a flattish spin the right way up with the undercarriage still down.The all-up-weight at take-off was 55,000 lbs compared to the maximum cleared weight of 70,000 lbs.

The aircraft appeared to hit the ground at an inclination of approximately 30 degrees with the starboard wing slightly lower than the port. The fuselage was completely disintegrated and burnt except the tail unit. The port wing had almost

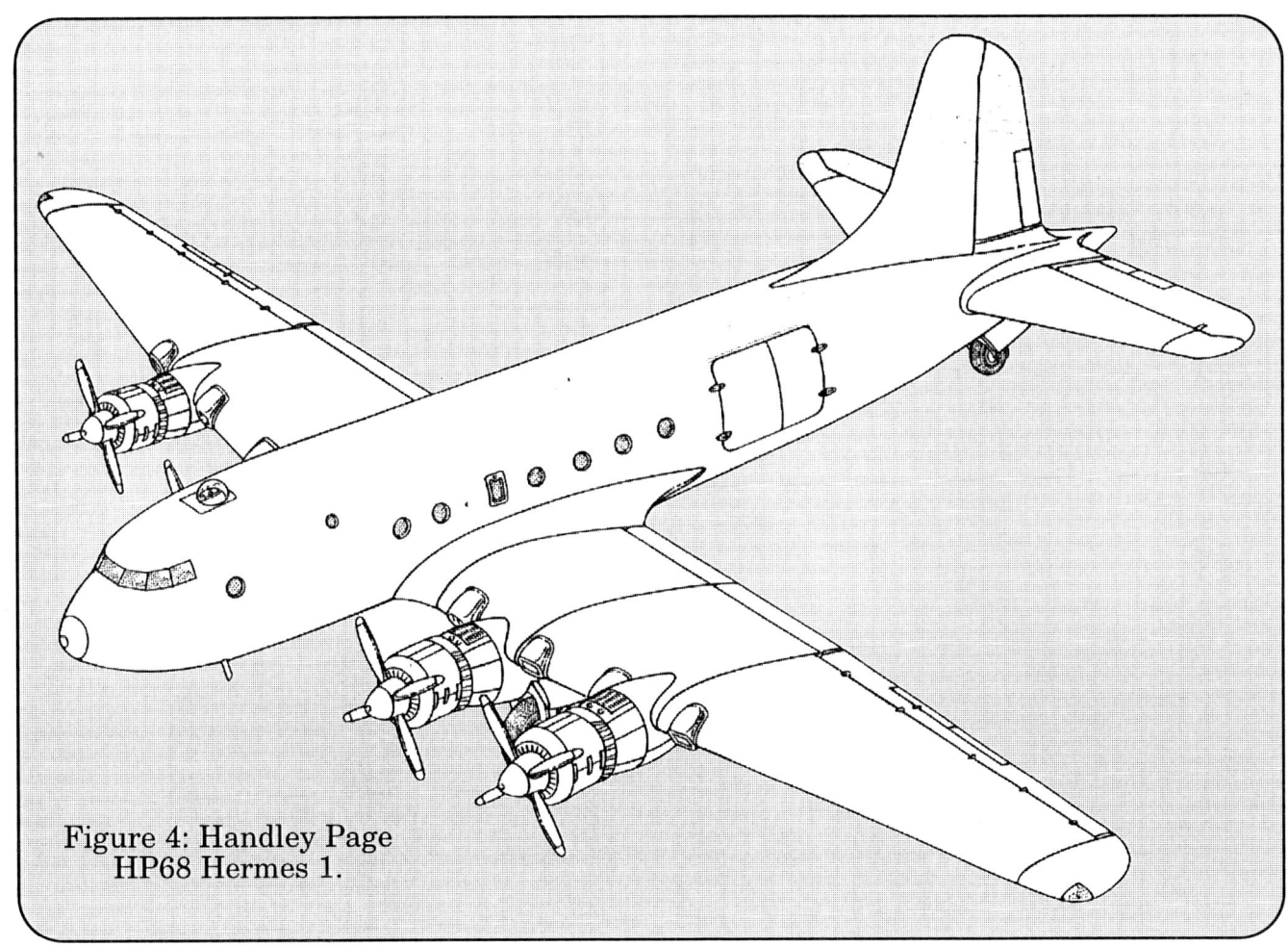

Figure 4: Handley Page HP68 Hermes 1.

HANDLEY PAGE HASTINGS AND HERMES

Hermes prototype G-AGSS at Radlett prior to the fatal test flight. The strong resemblence to the Hatsings is apparent, even to the freight doors. [Handley Page Assn]

completely disappeared, with pieces of it as far as 150 yards away, and the starboard propeller hubs buried about four feet down.

Investigation commenced immediately and on 6 December the report mentioned the fact that the film taken by Brailsford and Steel confirmed the observer's reports. The firm offered full co-operation and the MAP RTO at Handley Page stated that there had been full co-operation between the firm and the RAE, and that all the latter's suggestions regarding the fin and tail surfaces had been met by Handley Page.

A number of lines of enquiry were investigated, such as longitudinal snaking, elevator overbalance, elevator or tab control lines crossed, and locking of tabs or elevator. To check the elevator balance characteristics Handley Page were to repair the tail unit so that it could be tested in the RAE Farnborough 24 foot wind tunnel. In conjunction with this, Aero 3 section RAE Farnborough, were preparing to carry out calculations on longitudinal oscillations.

By 2 January 1946 the cause of the accident had still not been determined, although elevator overbalance was discovered during wind tunnel tests on the half tailplane of the first prototype, but this was between 10 degrees and the forces in this region were light. On 4 January Mr Lyons of the RAE Farnborough Aero Department stated that he considered that the pitching motions of the aircraft seen on the film were consistent with the pilot's attempts to overcome elevator instability, and that overbalancing of the elevators might make the aircraft extremely difficult to take-off and land, but would not cause the final stall and spin. The Director of RAE Farnborough wrote to the Secretary of State for Air on 16 January, concluding that the final cause of the accident remained a mystery, that the overbalance did not explain why the aircraft suddenly swooped and stalled. On 22nd of the month he also wrote to the Chief Inspector of Accidents to give a resume of the accident and examination carried out. In this was mentioned the failure of the elevator up-stop in a manner consistent with excessive elevator up movement.

The remarkable thing about the whole episode was recalled by E. N. Brailsford, who was a flight observer friend of 'Ginger' Wright. Apparently, a week or so before the crash Wright changed from his normal ebullient self to a morose man. It appeared he had experienced a most vivid dream in which he saw an aircraft crash, and rushing to the wreckage had pulled some-one from it - and found that it was himself! Noel Brailsford and Steel were involved in these flight tests and witnessed the effect of the dream had on their friend as well as the tragic fulfilment of his premonition.

Nevertheless trials and tests had to continue and on 27 August 1946 Hastings TG580 was flown to the AFEE for trials as a glider tug and for cooling tests in this role. After this it was returned on 28 September to Radlett for completion of the contractor's trials. Following this, between 17 January and 25 April 1947, the second prototype (TE583) arrived at Boscombe Down for a brief handling trial, being flown at an all-up-weight ,of 75,000 lbs, although the aircraft had originally been cleared to 70,000 lbs. The trial took place in two stages, first it was flown in the as received state, then modified to try and alleviate the faults found and then re-flown.

The first stage found the elevators too heavy, and stick free there was longitudinal instability in the climb with large changes of trim required with operation of the flaps. A change was made to the elevator balance gearing and a stabiliser spring (bungee) fitted, this brought the elevator control force and stick force nearly within limits, but it was considered that the control force at low cruising speed was unacceptable light. The stick free stability on the climb was low and also the change of trim with flap operation still remained unacceptably high. Also faulted was the minimum control speed for engine 'out' after take-off, which was considered high in relation to the stalling speed, Rudder tramping' had also occurred with full use of the rudder in certain asymmetric conditions, and the ailerons were lacking in response as well as being too heavy.

The next aircraft to join the testing circus was TG499, that on 11 August 1947 was dispatched to AFEE Beaulieu for paratrooping trials. This was followed by TG502 on 2 October being allocated to RAE for trials of radio and radar installations. TG502 was the second Hastings to make an official call at RAE Farnborough, for on 26 June 1946 TE580 had been flown there by Squadron Leader Hartford and demonstrated over three days.

The second Hastings to arrive at AWEE was TG503 and was the subject aircraft of the 2nd Part A&AEE 843 report. The test was primarily concerned with the behaviour of the airframe, engines and equipment during an intensive flying trial. This also included a flight to Australia, New Zealand and return, when during this trial and until the aircraft returned to Radlett on 5 June 1948 the flying time was 153 hours 20 minutes. Throughout this period it was found that in general the aircraft was reasonable easy to maintain; cabin heating, cooling and ventilation was adequate in flight and on the ground in temperate conditions, but inadequate on the ground in tropical conditions. No serious maintenance problems were encountered during the period away, and a number of pilots, including a retired bush pilot had flown the aircraft. As a result of the trial a number of modifications were considered necessary before the delivery of the Hastings to the Service. These included an improvement to the weatherproofing of the fuselage, cabin heating, and adequate ventilation and cooling under tropical conditions.

By now a number of different Hastings were being prepared at Radlett for various trials and roles, TG502 departing to Boscombe Down on 27 April 1948. TG506 departing on 29 September for winterisation trials at W.E.E in Canada. TE583 being prepared for development work on glider towing of a Hamilcar glider.

TG501 was the subject of the next part of the A&AEE report, having arrived at Boscombe Down in October 1948 for an intensive engineering and maintenance appraisal. No serious servicing problems were encountered and it was considered reasonably easy to maintain as a result of care and thought being given to maintenance requirements during the design stage. The only major problem found was the gaining of access to the back of the instruments on the pilot's panel. The aircraft was returned to Radlett for the incorporation of modifications as trial installations.

Hastings C1 TG501 was a production aircraft in which modifications required on the prototype had been incorporated, this was the subject of a brief handling test during October-November 1948, It was found that the change of trim with flap operation was substantially eliminated by the interconnection of the flaps and trim tab, and an improvement was also obtained by the reduction of the pilot's elevator trim gearing. Inadequate stick free longitudinal stability at low speeds, high stick forces in the landing approach and a change in trim with power 'on' were all criticised, while the need for a stall warning device was required.: Regarding the stall warning the firm tried various devices on the mainplane leading edges to promote pre-stall buffet, but eventually settled for a synthetic device to light a warning indicator lamp in the cockpit.

From 13 October to 9 November 1948 Hastings C1 TE580 was also at A&AEE, this was to determine the optimum flap setting for take-off. Tests were carried out to relate the flap settings and climb-away speeds to the minimum safe take-off distance with all engines operating. So the climb away was made at a number of different speeds, with the pilots asked to give what they considered the minimum safe speed to climb away, ignoring the possibility of engine failure. On this basis, at an all-up -weight of 75,000 lbs, it was determined that the flap setting was 40 degrees with a climb away speed of 100 knots IAS and a take-off distance of 1440 yards to clear a 50 foot barrier. The second method tried was to hold the aircraft down until the 'safe speed' was reached using a flap setting of 25 degrees. With

Hastings I TG506 modified for duty at the Parachute Testing Unit. [E.Watts]

HANDLEY PAGE HASTINGS AND HERMES

this a safety-speed of 119 knots IAS gave a take-off distance of 2100 yards to clear a 50 foot barrier.

To improve longitudinal stability TG502 had been fitted with its tailplane in a lowered position, and during May 1948 it had been at A&AEE for trials in this configuration, being the subject aircraft for Pt 8 of A&AEE /843 report. The report concluded from the trials that there was some improvement over its predecessor in regard to stability in the climb, but that unfortunately there was a deterioration in stability on a powered approach, aggravated by large trim changes and an increase in the stick forces on landing, and as such that it was not suitable for Service use. The same aircraft was the subject vehicle during July 1949 for checking take-offs with one engine inoperative and to determine whether a Hastings could be flown on three engines. A technique was developed where the aircraft took-off after a ground run of approximately 1100 yards on a level runway, and reached a height of 50 feet after a distance of 1470 yards with the critical starboard outer engine inoperative. When opening up the asymmetric engine however, it was essential that care was taken to prevent swing.

During this period, at AFEE was Hastings TE583 on trials towing a Hamilcar glider with checks being made of the various engine cooling modifications, and the cooling under ICAN and Temperate Summer conditions. The aircraft was fully test instrumented, including Sangamo-Weston pyrometry and an automatic observer. The trial determined that under tropical conditions, the Hastings at an all-up-weight of 66,000 lbs and the Hamilcar at a maximum weight of 37,500 lbs, the recommended take-off speed was 80 knots with a climb away speed of 120 knots. For ICAN and Temperate Summer conditions the Hastings could be flown at an all-up-weight of 68,000 lbs and the performance and cooling was satisfactory on take-off and climbs, but in level flight at maximum ECB condition, height could not be maintained. So it was laid down that the all-up-weight for towing a Hamilcar would be 66,000 lbs.

AFEE T57 report was issued on 29 September 1949 and covered the tests towing a Hamilcar glider, with the tests being carried out from Shaibah airfield in the Persian Gulf - and Service personnel who have been to Shaibah know how hot the tropical condition can be! The Hastings was TG533, and the Hamilcar was RZ427. The tug operated at an all-up-weight of 66,000 lbs and was a standard production aircraft, the combination taking 1225 yards for lift-off and 1925 yards to gain 50 feet. No specific difficulties were found when operating either the tug or glider under the tropical conditions.

Further trials were carried out at Boscombe Down and these covered handling trials at extended forward CG, handling with reduced tension in the control cables, as well as climb and level performance tests at a take-off weight of 78,000 lbs, The 16th Part of A&AEE 843 report dated 15 August 1950 covered a meteorological sortie from Gibraltar by TG503; the sortie was made under summer conditions at a take-off weight of 72,500 lbs. The sortie track distance of 1650 nautical miles was made in level flight and included a climb to 30,050 feet, and at the end of the task 20 percent of the initial fuel load still remained. This sortie determined that the performance, engine cooling and crew comfort on the Hastings was satisfactory for the task of meteorological survey.

Further trials on the Hastings covered a number of airborne support tasks, such as pannier dropping, containers and stores; as well as the external carriage and dropping of two 5 cwt Jeeps, six containers and a number of paratroops. TG500 was involved in these trials, one task being the carriage and dropping of one 5 cwt Jeep and a 75 mm howitzer, all these heavy loads were carried on the Heavy Stores Beam. As well as trials being carried out, a recommended procedure was established for the loading and dropping of the stores and paratroops. It was considered that provided the aircraft was modified to the current standard for tropical glider towing, then the operation of the Hastings in the heavy parachuting role could be safely carried out in tropical conditions. The take-off was normal using the flaps at maximum lift position, there was a tendency to swing to starboard, which became more active at higher take-off weights, but this could be held by coarse use of rudder and differential use of the throttles. It was recommended that for the dropping operation, the aircraft should be flown straight and level with the flaps in the maximum lift position,and the speed should not exceed 115 knots. On release of the stores there was no pronounced change of attitude and the slight nose up pitch could be easily held.

Between 27 January and 9 February 1949 handling trials were carried out on TG502, this was the first trial of a C2 aircraft. On this aircraft the tailplane and elevator area had been increased by 10 percent, the tailplane lowered 16 inches down the fuselage line from its original position, the tailplane incidence reduced by 2 degrees, a spring tab fitted to the elevator and a new nose shape built onto the elevator, These changes had come about due to the criticisms raised by the A&AEE on Hastings C1 TG501.

The results of the trial found the elevator control forces much reduced and the standard of longitudinal control at a satisfactory level. The rudder was considered rather ineffective at speeds below 90 knots,.but in normal flying conditions was in harmony with the elevator, both of them being light and effective, The main criticism was now levelled at the ailerons, which were still considered heavy and ineffective, Further development work on this problem was then made by the firm, who introduced modifications to lighten the aileron control and to eliminate the lateral oscillation which occurred in dives. The result of this was that`in June 1949, TG502 was given a brief handling trial to check on their effectiveness, Although not 100 percent right the

TESTING AND ALNALYSIS

Table 8 Hastings C2 take-off performance.

Type of Take-off	Start of climb away — Distance from rest	Start of climb away — speed at start	At 50 feet — Distance from rest	At 50 feet — Speed at start
At minimum practicable speed ignoring engine failure	1160 yds	100-103 kts IAS	1250 yds	100-102 kts IAS
At minimum practicable speed taking into account an engine failure.	1330 yds	105-109 kts IAS	1430 yds	105-107 Kts IAS
At safety speed quoted in Pilots Notes	1980 yds	125-130 kts IAS	2120 yds	125-129 Kts IAS

three controls were now in harmony at cruising speeds, but the rudder at low speeds was now lighter and less effective and at high speeds it was heavier.

The next major change affected the engines, this being the incorporation of the Hercules Mk.106 driving 13 feet diameter DH D100/446/1 type propellers. The take-off power of this engine had been uprated to 2800 rpm +10 psi boost. The Hastings take-off weight had been increased to 80,000 lbs and over April and June 1951 Hastings C2 WD476 was at Boscombe Down and the subject of the 6th Part of A&AEE843/1 report, The take-off performance was taken over three selected cases and was made with the flaps at the maximum lift position and the undercarriage in the 'down' position all the time. Listed in Table 8.

As part of the freight carrying development part of the Hastings' programme, with TG500 having done the test flying with the Heavy Stores Beam, the next step was the carriage of the Paratechnicon, which was a large freight carrier that was similar to the one flight tested on Halifax A7 PP350. The Paratechnicon was a large detachable freight container, that was designed basically for the carriage of vehicles, these would be driven in and then the container ends sealed off by large fairings. It was intended to be released in flight and to descend on parachutes, being first installed on C1 TG499 and flight tested for the first time on 13 May 1948.

On 26 September 1949 the combination took-off from Boscombe Down for a handling sortie to determine the effects that the Paratechnicon had on the directional stability of the Hastings. Approximately one hour after take-off, TG499 was seen at 4,000 feet approaching the airfield flying straight and level. Then suddenly the Paratechnicon was seen to have detached and the aircraft immediately climbed, the port wing dropped, and when the aircraft was nearly vertical it dived to port and eyewitnesses could see half of

Hastings I TG499 experimentally fitted with paratechnicon, on Boscombe Down trials for airborne support duties.
[E.Watts]

Table 9. Take-off distance under different conditions.

Type of Take-off	Total distance to 50 feet height		
	ICAN at S.L.	Tropical at S.L. (45C)	Tropical at 5000ft (35C)
At minimum practicable climb-away speed (105kts)	1500 yds	1980 yds	2450 yds
Climb away at minimum control speed for asymetric power (120 kts)	1880 yds	3120 yds	3700 yds
Climb away at safety speed from Pilots Notes (125 kts)	1970 yds	3550 yds	4200 yds

the starboard tailplane and elevator was missing. TG499 then commenced a series of five climbing and diving turns to port before crashing into high ground (Beacon Hill) near Boscombe Down.

Investigations determined that the Paratechnicon's nose fairing had disrupted under the air loads, initially along a longitudinal rivet seam. With the nose fairing gone the airflow into the container forced the Paratechnicon away from the aircraft, and in breaking away the Paratechnicon struck the starboard tailplane and elevator, the tailplane breaking away and consequently detaching the elevator. Wind tunnel tests and strength calculations carried out during the investigation determined that the loads in flight assumed during the design stage had been well under-estimated and a small degree of yaw above cruising speeds could cause failure of the nose fairing.

The Heavy Store Beam had been cleared for use in 1949 and between January and May 1952 Hastings C2 WD476 was at A&AEE on trials to check the performance and handling at 80,000 lb all-up-weight under ICAN and tropical conditions. Take-offs were made both at Boscombe Down and in the tropics, and, were made at a range of climb away speeds between 105 and 125 knots IAS, in order to assess, not only the minimum take-off distance, but also the normal take-off techniques, the results in Table 9

A number of trials were carried out on the Hastings C2 aircraft with various external loads and the dispatching of loads; then during June 1953, after a high rate of failures of the engine cooling fans, WD476 was dispatched to Boscombe Down to check the engine cooling with and without the cooling fans. The tests were carried out with the aircraft at an all-up-weight at take- off of 80,000 lbs, and determined that the cylinder head temperatures could be exeeded under Tropical Summer conditions if the cooling fans were removed, As against this result, due to the failures of the cooling fans in Transport Command, Hastings aircraft were being flown without them, but there had been no undue increase in engine or component failures, so it was suggested that the cylinder head temperature limitations for the Hercules 106 could be materially exceeded with safety.

Various trials with the Hastings were carried out over the years, but these were usually related to the testing of equipment, dropping of different types of stores, and flight development of radio and radar installations. During these years of testing, the personnel of A&AEE; AFEE and RAE as well as Handley Page, had the aim of producing a viable transport aircraft for the RAF, which could be operated in many roles and climates. In spite of incidents and accidents, this aim was pursued and the Hastings emerged as a rugged versatile aircraft.

Hastings on trials, loaded with 5 cwt Jeep, 75 MM Howitzer and supply containers on heavy stores beam. [RAE]

HANDLEY PAGE HASTINGS AND HERMES
Chapter Five
The Hastings aircraft

Fuselage construction.
This is constructed in two sections (but after manufacture remains as one unit), with bulkheads at frames 240 and 820. The structure, like most Handley Page designs, was strong, with a circular cross-section of 11 feet diameter. It was skinned all over with aluminium alloy sheet attached to longitudinal stringers, the latter attached to the outer flanges of the frames. Most of the frames were of 'Z' section, but at frames 330 and 420 (spar pick-up points) the frames were heavier and consisted of extruded light alloy flanges with sheet metal webs. All the frames were numbered, representing inches from the nose of the aircraft. Stabilisers (intercostals) were fitted between the frames to give support, with top hat stringers running from the nose to the tail (drawing 4).

There were four main doors provided in the fuselage, and at the freight door position box section frames were fitted, with angle section longerons above and below the freight door and paratroop door. On the C4 aircraft there was no paratroop door and the freight door was replaced with a normal passenger door and power operated steps (drawing 8).

The fuselage floor consisted of Plymax panels, which were riveted to a structure of transverse plate beams, that was further supplemented (so as to provide a stronger load bearing surface for wheeled vehicles) by the fitting of intercostals at a closer pitch at the outboard end of the floor than in the centre, with stronger strut support underneath the floor (drawing 5). Shallow longitudinal channels for the accommodation of seat fixtures and also lashing points were recessed into the floor.

The flight deck provides for two pilots, navigator, flight engineer and radio operator, the pilots having dual control as standard. The main controls, such as the throttles and rpm levers, were neatly grouped together on a central pedestal that was convenient for both pilots. Control of the fuel system and various engine and airframe controls was by the flight engineer, who also had duplicated throttle and rpm levers. The crew compartment was separated from the main cabin by a wooden bulkhead at frame 240.

The bulkhead at frame 820 consisted of a light alloy web plate with lipped angle and 'Z' section stiffeners. Two boxlike booms of built-up construction were riveted to the bulkhead and by means of lugs provided attachment for the fin main spar.

The passenger cabin has an internal diameter which averaged 10 feet 4 inches, with a fuselage outer diameter of 11 feet, the size of which had been determined for the HP64 in 1943 wind tunnel tests. Cabin sections were pressurised at the works during the design stage,

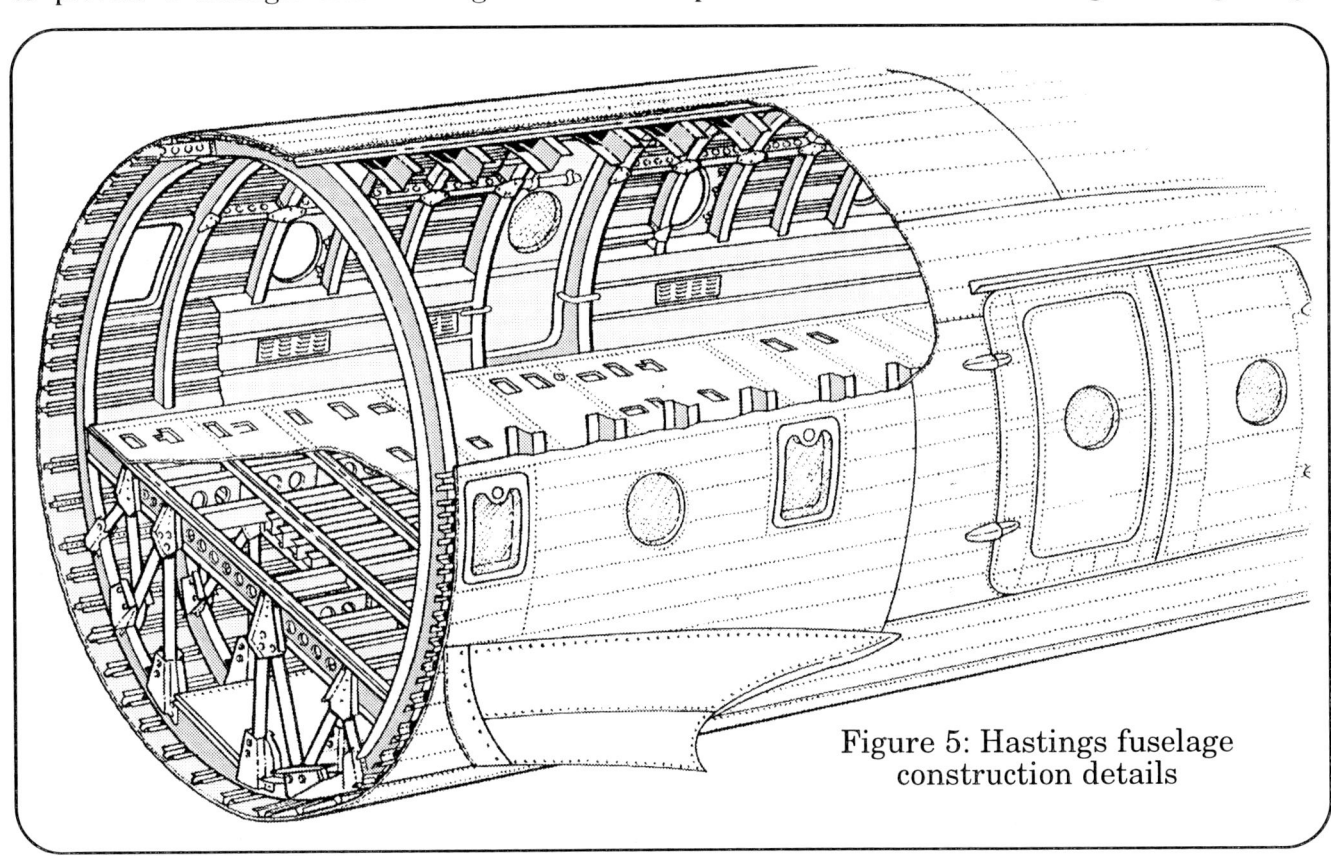

Figure 5: Hastings fuselage construction details

possibly with a view to later sales of a pressurised civil design.

Escape hatches and portholes were provided each side of the fuselage and an astro-dome was fitted in the fuselage roof above the gangway adjacent to the navigator's position; the assembly was detachable and fitted between adjacent frames. Downward observation windows were provided in the nose for supply dropping, and were fitted below the pilot's floor in the fuselage bottom skin between frames 6.42 and 5. The pilot's windscreen and canopy consisted of an upper and lower light alloy frame with vertical members into which were fitted the transparent material. Fitted in the corner window of each pilot's windscreen was a DV panel, and aft of each DV panel was fitted a sliding side window.

The freight doors of the C Mk.1 and C Mk.2 were of similar construction, with each comprised of a box-section frame that was externally skinned with light alloy sheet and internally stiffened with top-hat section stringers.

Tail unit construction.
This consisted of a conventional tailplane and elevators with a fin and rudder. Each tailplane and elevator was constructed in halves, port and starboard. The tailplane was constructed around a main spar with a fore and aft drag member, the whole covered with a stressed metal skin that was riveted to those members and the pressed metal ribs. The tailplane main spar was comprised of a light alloy sheet web that was riveted to 'T' section metal booms (drawing 6).

The elevators consisted of a light alloy plate web 'D' nose to which was attached pressed sheet tail ribs, all covered with a stressed metal skin. A mass balance mild steel plate was riveted inside the leading edge of each half elevator. Also on each elevator was mounted a balance tab outboard and a trim tab inboard.

The fin construction was similar to the tailplane, with the lower part of the nose extending forward to mate with a small dorsal fin which was constructed integral with the fuselage. The rudder was similar in construction to the elevator and on it was mounted a spring tab and balance tab. It should be noted that there were dimensional differences between the tailplane and elevators of the prototype and the C1 and C2, and also the elevator nose shape.

Mainplane construction.
These, though inherited from the projected HP66 bomber, were basically Halifax in regard to the two spar construction and distance between spars. Their main difference lay in the extension outwards of the intermediate plane so as to incorporate the outer engine nacelles, these on the Halifax had been mounted on the outer plane.

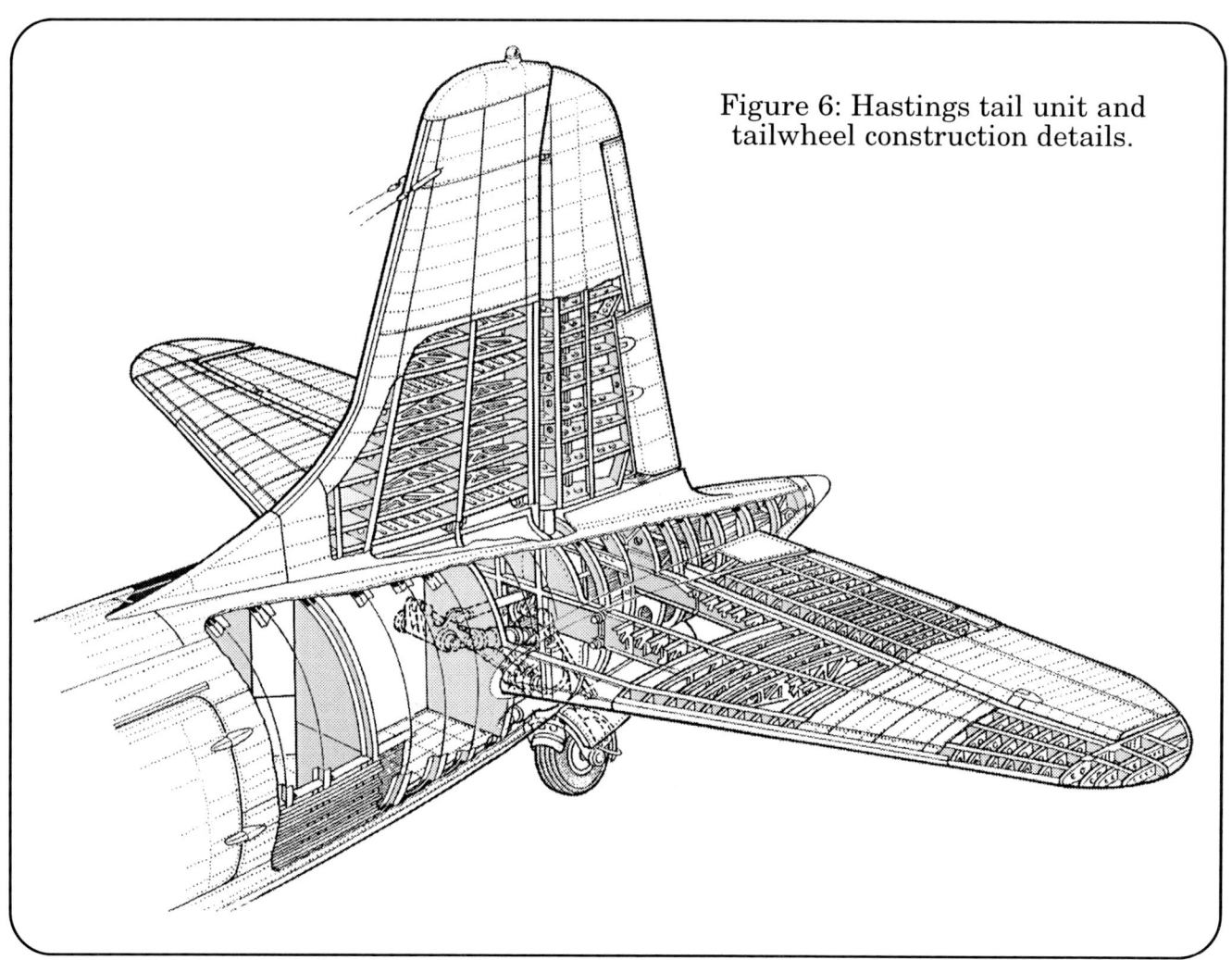

Figure 6: Hastings tail unit and tailwheel construction details.

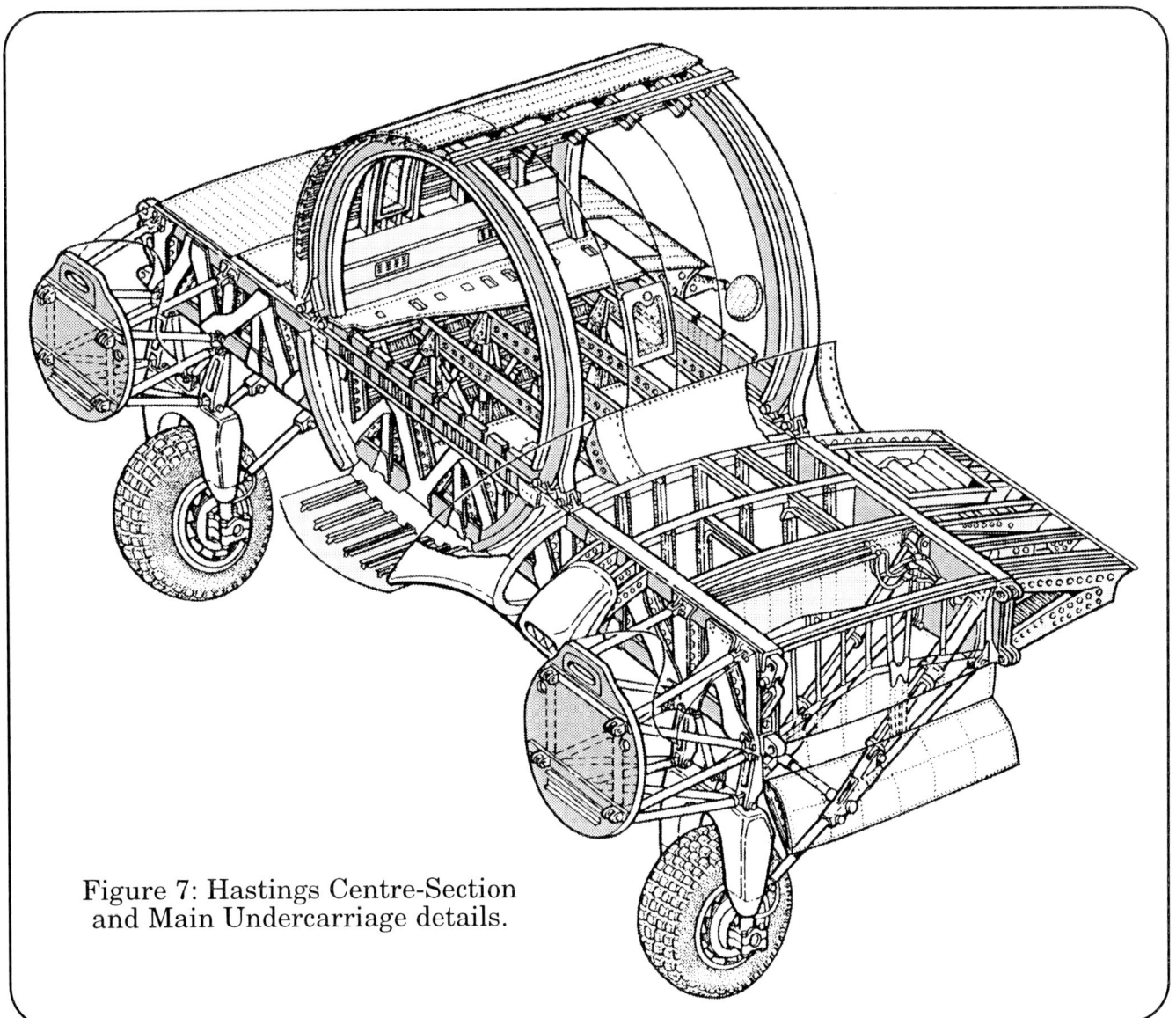

Figure 7: Hastings Centre-Section and Main Undercarriage details.

The mainplanes were of all metal construction and consisted of five sections, a centre-section integral with the centre fuselage, intermediate planes (port and starboard) and outer planes (port and starboard). Each wing section was built on a two spar construction, with the wing to wing joints consisting of high tensile steel fork ends and eye fittings on the top and bottom booms, these were secured by special high tensile steel joint pins that were locked by caps and split-pinned nuts and bolts. The spars in the centre-section were attached to the fuselage frames at frame number 330 for the front spar and frame 420 for the rear spar. The centre-section front spar is constructed of built-up aluminium alloy extruded channel booms that were braced with box-section plate diagonals. The rear spar was constructed of 'T' section extruded booms with a sheet metal web braced by vertical angles (drawing 7).

The aerofoil section of the mainplanes was as for the Halifax, being NACA 23021 for the centre-section and intermediate plane, and NACA 23009 for the outer planes. The wings, outboard of the centre section, taper outboard in chord and thickness. The front and rear spars, as with the Halifax, differed throughout their length, with the ribs also of different construction, some being of plate with others of built-up box booms and plate webs. In the intermediate wings, the main spars consisted of a front spar with 'T' section booms riveted to a sheet metal web and a rear spar of angle section booms riveted to a sheet metal web. A false spar comprised of a sheet metal web and extruded angle booms was fitted in the intermediate wing forward of the front spar, and extended between the two air intakes in the leading edge of the wing.

In the outer wings the construction of the front and rear spar was comprised of a sheet metal web riveted to extruded angle booms. On the outer wings were mounted the Frise type ailerons, that extended from the inboard end of the outer wing to the wingtip joint. Each aileron was constructed of a 'D' box spar, flanged plate ribs, and like the remainder of the win was skinned with alclad sheet. Each aileron was mounted on five aileron hinges on the outer wing. Each aileron carried a spring tab and a pilot controlled trim tab.

Four flaps were fitted and were of the slotted version with two hinged to the centre-section and two on the intermediate wing (one each side).

HANDLEY PAGE HASTINGS AND HERMES

These were constructed around a light alloy box spar with its nose portion supported by plate diaphragms and tail ribs to the rear, the whole metal skinned.

All the fuel tanks were mounted between the wing spars, with tanks 1 to 5 mounted on bearers secured to brackets on the ribs and the rear spar. Tank 6, consisted of three flexible tanks, was supported by a light alloy flooring riveted to stringers extending to and attached to the ribs, and secured in position by lacing to the rear spar and baffles.

The nosing of the outer wings formed a complete unit, with the ribs bolted to brackets on the main front spar top and bottom booms, the ribs being of pressed flanged plate stiffened by fluting. The top skin of the nosing was riveted to the skin of the outer plane, which projected beyond the top of the main spar. Along the leading edge, flanged fittings were fitted to accommodate the TKS de-icing strip elements. The tail box of the mainplane was also a separate component, and consisted of ribs made up of upper and lower booms of 'Z' section, that were attached to brackets on the rear spar, and at certain points along the box were reinforced, the whole skinned over and riveted.

Undercarriage.

This was comprised of two main wheel units and a tailwheel unit, all of which retracted to the rear by hydraulic jacks, and on retracting, by means of linkage, closed the wheel-doors over the wheels. Each main unit was based on a magnesium casting housing two oleo-pneumatic shock absorbers working in sliding tubes, all of Electro Hydraulic manufacture. The arch was hinged to the bottom of the inboard engine mounting superstructure, with the single wheel and axle bolted to the bottom of the shock absorber assemblies. The retracting jacks were hinged to the top boom of the rear spar with their piston rod ends attached to the lower fittings of the upper radius arms, which had their upper ends attached to the lower boom of the rear spar. The lower radius arms were pivoted on the lower part of the undercarriage arch.

The tailwheel unit was retractable and was comprised of an upper and lower casting linked by a compression strut, that was retracted by an hydraulic jack. The design of the shock absorber allowed for free movement over rough ground, whilst damping out displacements created by landing shock loads. The tailwheel fork had a slight angled trail to ensure castoring action when taxying. A hydraulically operated locking device was also fitted that could be engaged by the pilot for take-off.

The mainwheel units were locked in the 'up' and 'down' position by internal mechanical locks that were incorporated in each hydraulic jack, each main wheel had pneumatically operated brakes. In the 'up' and 'down' positions, the main unit mechanical locks were prevented from disengagement by back pressure in the return line maintained by a non-return valve. Failure of either the engine or the hydraulic pump was covered by the use of an emergency hand pump for hydraulic operation. Failing that, an emergency air system could be used that admitted compressed air into the top of the jacks to force the undercarriage down and locked. This latter use necessitated the bleeding of the hydraulic system afterwards. The difference from this and the Messier system on the

Starboard undercarriage and inboard nacelle of Hastings T5. [Author]

THE HASTINGS AIRCRAFT

Halifax, was on the Halifax compressed air forced down separators in accumulators and forced down the undercarriage and flaps through the medium of oil when the system(s) were selected. The accumulators were isolated by stop-cocks after 'up' operation.

Hydraulic system.
The hydraulic services obtained their power from two Lockheed VI pumps mounted on the inboard engines, which drew their fluid from a 42 gallon reservoir. On pre-mod 756 aircraft an electrically driven booster pump in the reservoir was controlled by a circuit breaker on the flight engineer's panel. A main hydraulic accumulator with a separator was incorporated in the main circuit to maintain constant pressure and to damp out pump fluctuations; this was charged to an air pressure of 1600 psi. An accumulator was used to lower the flaps in the time honoured Messier manner, hydraulic pressure being used to raise the flaps. A compressed air system was used for the emergency lowering of the undercarriage and could be used for lowering the flaps as well. When mod' 1308 was incorporated then an additional air bottle was fitted for undercarriage lowering.

During normal hydraulic operation, power was supplied from the pumps through a filter to an automatic cut-out and on to the service selector. On completion of the operation, back pressure operates the cut-out and pressure oil circulates back to the reservoir. This means that the operation is complete and pressure oil circulating back to the reservoir with both pumps on an idling circuit still pumping.

Flying controls.
The flying controls were all quite conventional in their operation, with the control's trim tabs controlled by handwheels conveniently situated on the control pedestal. These latter control tabs were operated through the medium of cables and chains. The rudder and elevator surfaces were operated through push-pull tubes and the aileron surfaces through pushpull tubes, cables and chains. The elevator control was through fore and aft movement of the control column, and through pushpull tubes to the elevator torque tube, where a pre-load unit was fitted to give 'feel' during initial operation of the control. A spring bias unit was fitted between frame 934 and a lug on the starboard torque tube flange to improve the longitudinal stability of the aircraft.

At the bottom of each rudder bar post was a lever, which by means of a tube connected the two rudder bars. From the port lever a further lever set at 90 degrees transmitted movement through the push-pull tubes to the rudder quadrant at the base of the rudder, and the quadrant through the medium of a spring tab mechanism transmitted the movement to the rudder.

The aileron's movement was controlled by the turning of the pilot's hand-wheel at the top of the control column, this through a chain-wheel assembly transmitted motion to the main cable - run. At frame 390 the cables passed through pulleys and a channel drive on the face of the rear spar, where they connected with the aileron push-pull tubes. These at their outboard ends connected through a bell-crank to the aileron spring tab torsion bar and to the aileron itself.

Flap control was by means of a lever mounted on the pilot's control pedestal, this operated the flaps through the medium of a hydraulic jack mounted aft of the rear spar, each end of which was connected to a master lever. This was connected through an adjustable push-pull rod linked to the centre plane flap, that was linked to the other flaps by linkage and interconnected by balance cables. The flap lever operated in a gated slot to give four flap positions, UP, INTERMEDIATE, MAX' LIFT and DOWN.

Fuel system.
The Hastings fuel system was, like the Halifax, based on a number of fuel tanks, with groups of tanks feeding to a collector tank. There were seven tanks each side with the number 1 tank positioned inboard of the inboard engine. The tanks 2 to 5 were mounted in the intermediate wing with tanks 6 and 7 mounted in the outer plane. Electrically operated immersed fuel pumps were mounted in the number 1 tank and transferred fuel through non-return-valves (NRV) to number 4 tank, while the same type of fuel pump was fitted in the number 2 and 5 collector portions of those tanks to augment the engine driven fuel pumps.

Each group of fuel tanks supply one engine independently, or can be connected by crossfeed and/or balance cocks to supply one or more engines. Fuel from number 1 tank transfers to number 4 tank, and tanks 3 and 4 feed through NRVs to the collector tank in number 2 tank. Similarly, tanks 6 and 7 fed through NRVs to the collector tank in number 5 tank. The tanks were not self-sealing and were constructed of light alloy plates joined by De Burgue rivets. Tank number 6 was comprised of three lightweight Marflex flexible bag-type tanks, listed as 6A, 6B and 6C, these were interconnected and so functioned as one and were considered as one for use. To increase range two auxiliary external tanks could be installed, these were non-jettisonable, and when installed on the Mk.1 make it a Mk.1A aircraft.

In fuel management, it was recommended to fill the outer fuel tanks first and to use them last, as this improved the main spar life. For fuel jettisoning, controls were situated on the panel above the flight engineer's centre panel and three levers permitted the jettisoning of the fuel, enabling 2290 gallons of fuel to be jettisoned.

De-icing system.
The wing leading-edge, tailplane leading-edge and fin leading-edge were all fitted with TKS de-icing equipment, with the system pump fitted in the centre-section. The pump operation was controlled by an electrical control on the flight engineer's panel and the pump drew fluid from a

HANDLEY PAGE HASTINGS AND HERMES

44 gallon tank positioned in the inboard end of the port centre-plane.

For propeller de-icing two pumps were provided in the starboard side of the centre-section, and drew fluid from a 44 gallon tank installed in the starboard side of the centre-plane. Each pump had two delivery outlets and served two engines, all being controlled electrically from the flight engineer's panel.

The windscreen anti-icing was self-contained and operated by a hand pump on the starboard side of the second pilot's position.

Mk.C4 differences.

The floor structure of the C4 was similar to the standard Hastings, except that certain fore-and aft intercostals were omitted, as the VIP version did not have the same floor strength requirements. There was only the main passenger entrance door, and that was hinged downwards from its lower edge on the fuselage. The door incorporated in its structure a set of light alloy folding stairs, which extended and retracted with the door operation (drawing 9). Both the door operation, and the extension and retraction of the stairs was by means of a self-contained hydraulic system that was controlled by the quartermaster.

The floor was of Plymax panels, but as distinct from the C2, no expendable plywood covering was provided. The general seating, furnishing and equipment probably made the C4 one of the most luxurious standard VIP aircraft of the time, except for the lack of pressurisation, as it had sleeping booths, luxury lounge, washing and toilet facilities - or as one crew member remarked *"Everything plus"*.

Powerplant installation.

This was comprised of an engine mounting constructed of steel tubing and forged end fittings. It was secured at the inner end to the nacelle structure, which was also of tubular steel members and forged end fitting. The whole was attached to the mainplane at the front main spar. Forward of this structure was the fireproof bulkhead, on which was mounted the various services and control dis-connection points.

Around the inner engine nacelle the ribs were reinforced to take the loads from the undercarriage as well as the engine loads. All the cowlings of the inner and outer powerplants were removable. On the inner engine nacelles the undercarriage doors were attached to the nacelle skirt fairings, which were also removable, as were the rear fairings. On the outer nacelles all fairings were removable.

Each engine was cowled with aluminium or aluminium alloy sheet cowlings, with controllable gills operated by the flight engineer through electrical controls, so as to control the amount of cooling air. Air intakes were provided with ducting to all engines, and oil cooler and air ducting to the accessories. One of these air intakes provided RAM, CLEAN and WARM air to the fuel injector, selection being by the operation of actuators selected by the flight engineer.

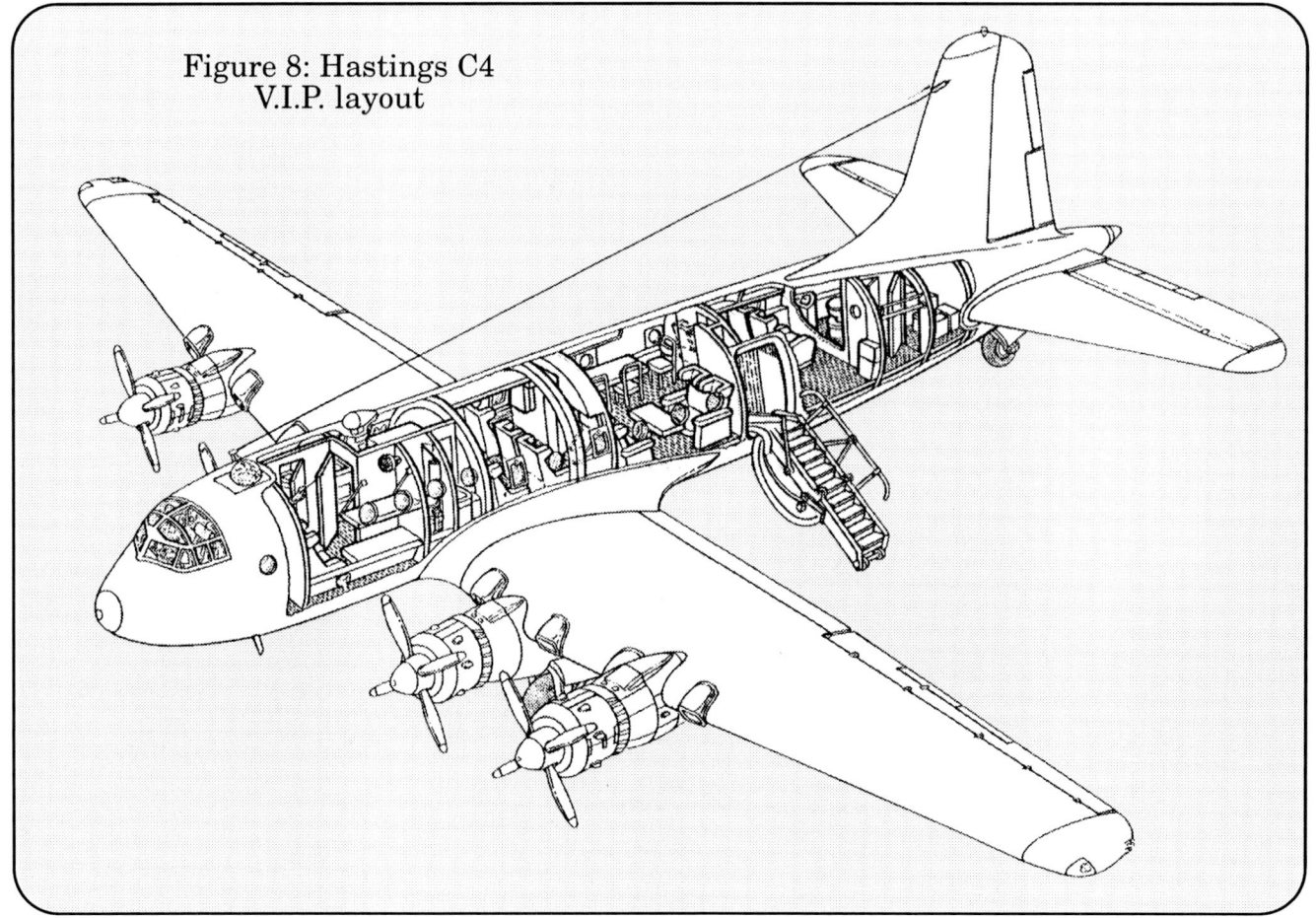

Figure 8: Hastings C4 V.I.P. layout

Figure 9: Hastings C4 on-board airstairs

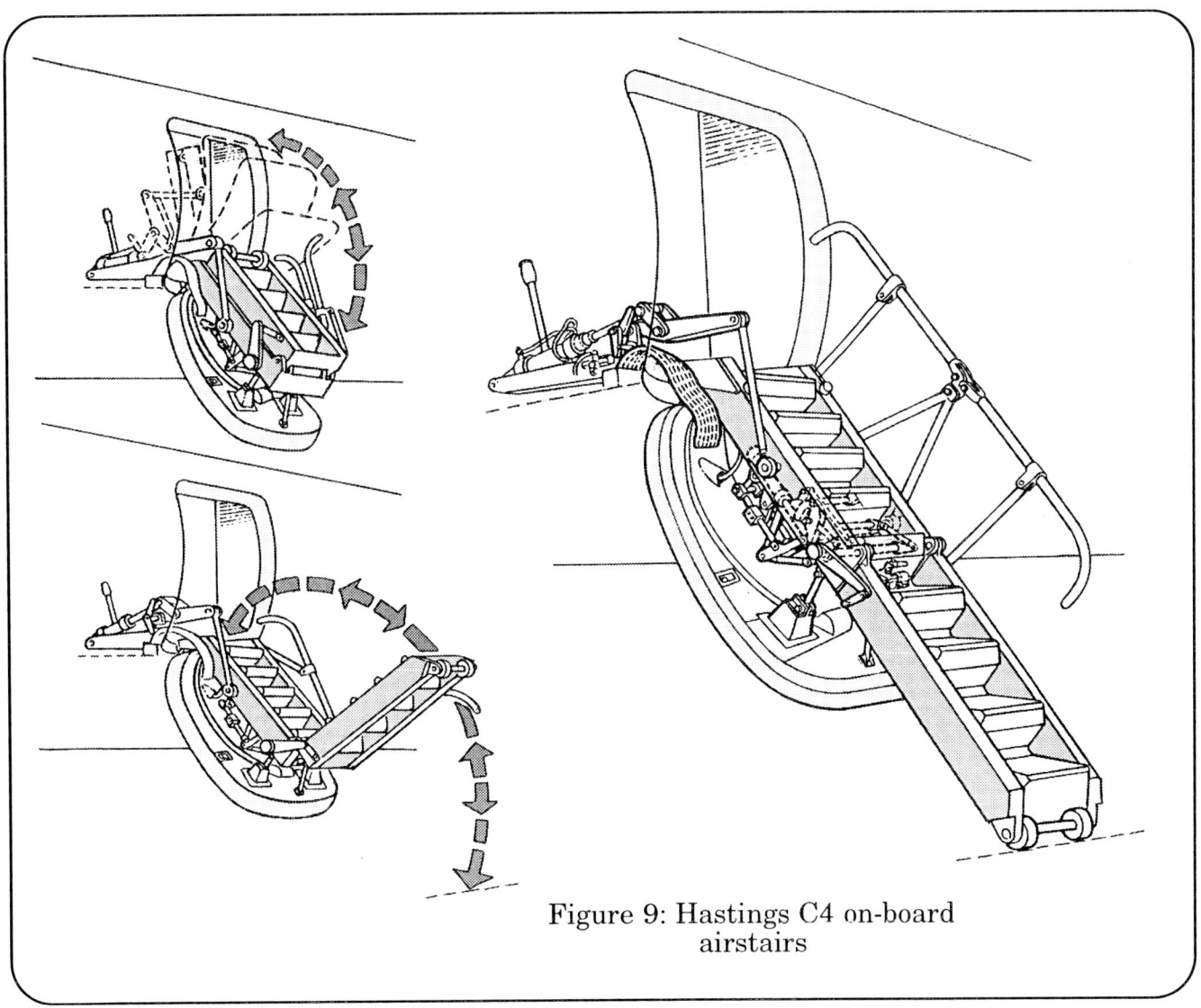

An auxiliary gearbox was mounted on the aft face of the engine bulkhead. The exhaust gases were taken to atmosphere by radially arranged exhaust stubs that ejected direct to the rear, with two that extended aft and passed through heat exchangers, these supplied hot air to the cabin air supply system. Accessories mounted on the gearboxes were:-

Inboard engines
suction pump
hydraulic pump
air compressor
type P generator

Outboard engines
suction pump
(starboard only)
air compressor
(port only)
type P generator

The engines powering the Hastings and Hermes were the Bristol Hercules, a progressive and direct development of those that had powered the Hastings' predecessor, the Halifax. The Bristol Hercules was one of the Bristol aero engine family of sleeve valve engines, that had been the first major and successful break-away from the standard reciprocating piston engine layout since World War 1, in that they used a single-sleeve valve per cylinder to replace the poppet valve gear of other makes of engine. The sleeve valve required no adjustments in service, so this reduced maintenance and the number of working parts, as well as ensuring constant and correct valve opening and closing, irrespective of engine hours run. In deleting the overhead poppet valve gear the cylinder head (known as the junkhead) allowed, during the design stage, the positioning of the spark plugs in the optimum position for efficient combustion.

In the Hastings the Hercules engines were

Below: Hercules engine nacelle of a Hastings T.5 [Author]

Table 10 Basic data of Hercules 100 series engines.

Number of cylinders	14
Arrangement of cylinders	two row, air cooled
Bore & stroke	5.75 x 6.5 inches
Swept volume	2360 cubic inches
Compression ratio	7.0 to 1
Supercharger gear ratio MS	6.679 to 1
FS	8.365 to 1
Reduction gear ratio	0.444 to 1
Fuel grade	100/130 octane
Weight, nett dry including mounting ring	2045 lbs
Take-off power at Sea Level	1675 hp at 2800 rpm +8 1/4 psi
Maximum power rating	MS 1800 hp at 2800 rpm at 9,000 ft
	FS 1625 hp at 2800 rpm at 19,500 ft
Max cruising power	MS 1215 hp at 2400 rpm at 12,250 ft
	FS 1125 hp at 2400 rpm at 21,000 ft

initially the Mk.101 and 106 and were accommodated in the 'E' type powerplant, which was a development of the RAE Low Drag Powerplant for the Hercules 100 engine. The intention was to reduce the internal cooling drag, reduce the mean cylinder head temperatures and reduce the overall drag. After bench tests at RAE Farnborough the powerplant was flight tested on Halifax RG642 so as to carry out a series of correlation tests at various powers, air speeds and altitudes. Although there was a discrepancy between the bench and flight tests, the final development was a good improvement over the standard model.

The original Bristol sleeve valve engine was brought to fruition after five years of research with the expenditure of £2,000,000. With this and further improvements the Hercules had been developed into the Mk.100 (drawing 10), which had an improved turbine entry supercharger, improved induction system, Hobson RAE fuel injector and an oil system with a constant level device. The Hercules 101 and 106 installed in the early Hastings aircraft were two-speed supercharged models, but in later aircraft the Hercules 216 was introduced.

Hercules 216.

This was basically the same engine as previous Hercules engines, but it incorporated a number of improvements based on operational experience and with the two-speed supercharger replaced with a single-stage. The supercharger gear ratio being approximately midway between the previous MS and FS gears.

The Mk.216 superceded the Mk.106 in most refurbished Hastings as well as the C2 and C4 aircraft, and so provided an improvement in operating and reliability.

The powerplant was provided with a comprehensive fire protection system which, triggered by the operation of the feathering button, closed the cowl gills, engine cocks and oil cocks, and after a delay operated the Graviner fire extinguisher bottles. These were also operated in

Table 11. Basic data on Hercules Mk.216.

Bore & stroke	5.75 x 6.5 inches
Swept volume	2360 cubic inches
Compression ratio	7.0 to 1
Reduction gear ratio	0.44 to 1
Supercharger gear ratio	7.26 to 1
Carburation	Hobson RAE BI/BH17
Fuel	100/130 octane
Weight, nett dry including mounting ring	2205 lbs
Maximum take-off power	1800 hp at 2800 rpm +12l psi at S.L
Maximum power rating	1925 hp at 2800 rpm at 8750 feet
Maximum cruise rating	1150 hp at 2400 rpm at 15,750 feet

Figure 10:
Bristol Hercules 100 power unit.

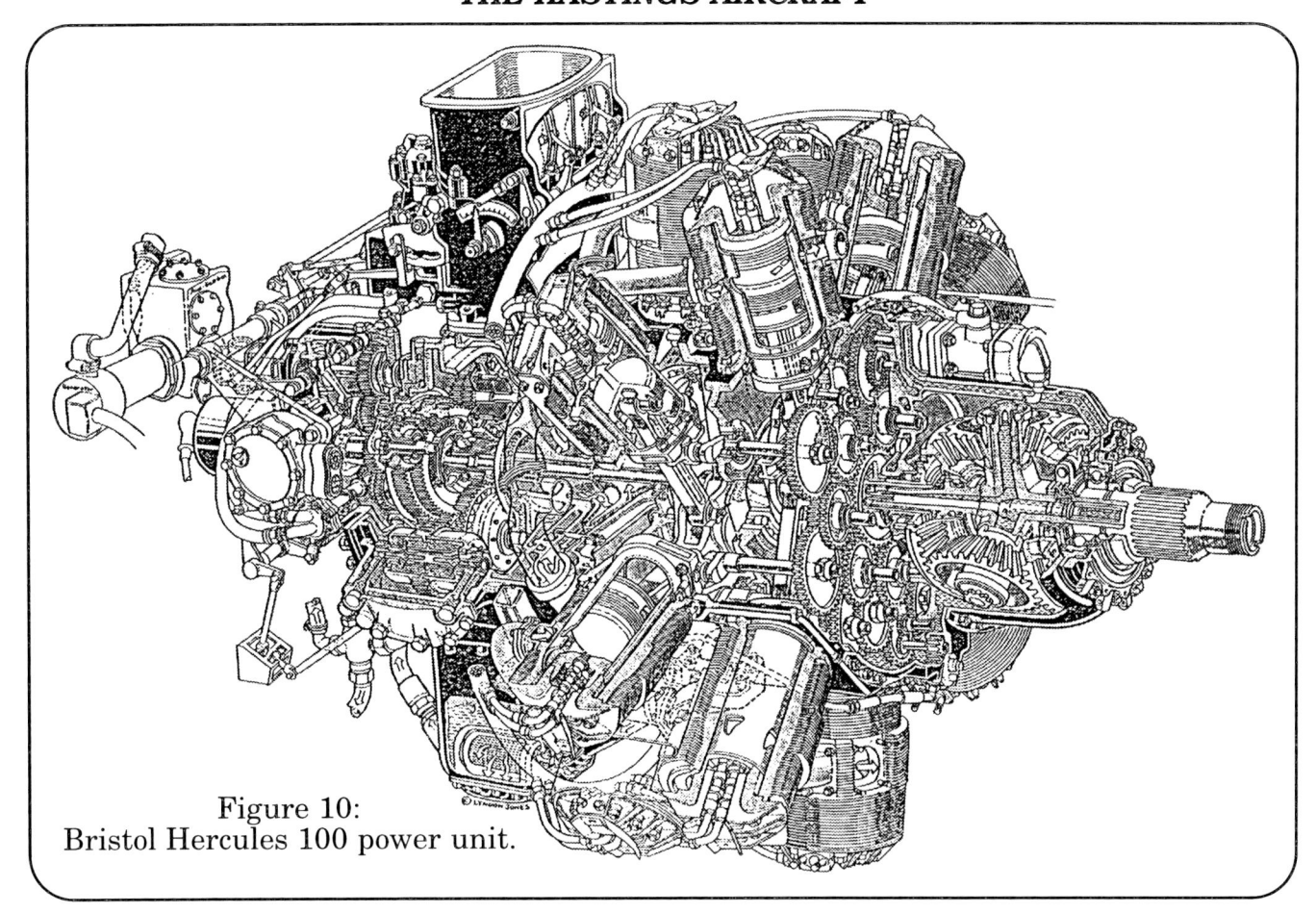

the event of a crash landing. The system being based on the wartime work of the RAE Fire Panel.

Handling. Flying controls.
The controls were effective throughout the whole cruising range, but the aileron response was not so good at low speeds below 120 knots (138 mph). The rudder was light and powerful throughout the speed range and it was therefore advisable to be used with care.

Under certain conditions of flight, rudder overbalance could be encountered, for if, when flying at low speeds with power settings in the rich mixture range, coarse rudder was applied and the wings were held level by opposite aileron, rudder 'tramping' could develop. The intensity increased with increasing angle of sideslip. If this condition was allowed to persist, rudder overbalance could occur and a very heavy foot force was required to overcome it, the rudder trimmers were recommended not to be used. Such conditions would not normally be encountered, except possibly when in asymmetric flight.

If rudder 'tramping' occurred in asymmetric flight, it was possible it would be the approach of the critical speed; an immediate increase in speed could be sufficient to reduce the 'tramping' and so avoid the possibility of sudden rudder overbalance. Alternatively, if the speed was above the appropriate safety speed, then an immediate reduction in rudder or aileron angle needed to be made. If slight bank was applied away from the failed engine, adequate control could be maintained in asymmetric flight, without the risk of approaching the critical sideslip angle.

The trimming tabs on all the three flying controls were powerful and sensitive, and the elevator trimmer in particular had to be used with care.

Flight handling.
On take-off the aircraft was aligned up to ensure the tailwheel was straight, then tailwheel lock was engaged. The brakes were released and the throttle levers opened up smoothly to the take-off position. There was a tendency to swing, but the rudder was effective enough to correct any small deviations as soon as the tail was up. The tail rose of its own accord at 60-70 knots, even with the CG on its aft limit. A light steady pull force could be used to unstick at the correct speed for weight, see table 12. The maximum crosswind component for take-off was 18 knots, but with external tanks on the Mk.1 it was recommended that this was limited to 14 knots. Once clear of the ground the undercarriage was retracted and once at 300 feet AGL then the flaps could be selected up. The climb was continued at 150 knots.

When established on the climb the engine power was reduced to 2200 rpm. The aircraft trim changed with the undercarriage 'up' to nose up, and with the undercarriage going down to the 'down' position then the trim changed to initially nose up and then nose down. Flap movement from the 'up' to 'quarter down' gave nose up trim, from quarter to maximum lift position gave nose up

HANDLEY PAGE HASTINGS AND HERMES

Royal Aircraft Establishment low-drag Hercules 100 powerplant on test on Halifax RG642 at Farnborough. [RAE]

trim. From maximum lift to 'down' the trim was slightly nose down.

The stalling characteristics of the Hastings can be summarised as follows. The stalling speed varied with all-up-weight, and with power 'off' and the undercarriage and flaps up (stalling speed 108-122 knots at 65,000-83,000 lbs), buffet could be noticed just before the stall. At the stall, buffeting increased and the nose dropped gently, but there was no tendency for either wing to drop. With power 'off' and undercarriage and flaps down (stalling speed 90-102 knots at 65,000-83,000 lbs) the behaviour was similar, but there was less warning of the approach to the stall and nose drop could be greater. With power 'on' and undercarriage down (stalling speed 80-85 knots at 65,000-74,000 lbs) there was no warning of the approach to the stall, the stall being accompanied by marked aileron snatch and wing drop, either wing could drop. When the wing had dropped then sufficient rudder to prevent yaw had to be applied and the ailerons held neutral; normal recovery action was effective, although considerable height could be lost in recovery.

If an engine failed on take-off, at or above the safety speed, the aircraft could be held straight and would maintain height while the undercarriage was retracted and the failed engine's propeller feathered. With an outer engine failure the foot load on the rudder was heavy and care had to be taken not to overcontrol, or rudder 'tramping' and overbalance could occur. Rather optimistically the Pilot's Notes stated *"slight tendency to swing on take-off"*, but an analysis of Hastings pilot's remarks gives a different picture, far from slight the impression is given that the swing was strong, although it could be held by normal action.

Most pilots who regularly flew the Hastings came to love the beast, but it could bite on landing, and many a pilot who became complacent got bitten - although one joker did remark *"It was built to fly after all, not sit on the ground"*. The control cabin was well laid out and the pilot's seats were some of the finest ever fitted to a military aircraft and set a standard for years to come, being a boon on long-distance flights; they could be adjusted in all directions and were complete with arm and headrests.

The Hastings appeared to remain serviceable when operated at a high frequency, and this is noticeable when checking operations or services where the aircraft was operated continually, as the defect numbers or unserviceability was then remarkably low.

Hastings CMk.4 WD500 VIP transport. The only real visible difference is the lack of freight doors [IWM]

HANDLEY PAGE HASTINGS AND HERMES
Chapter Six
Service use of the Hastings

Introduction

The Hastings' introduction into service was to be, like its predecessor the Halifax, "slung in at the deep end", for its service trials began on the Berlin Air Lift in November 1948, operating along civilianised Halifaxes. From then until October 1949 (on completion of the operation) the Hastings carried out 12,396 sorties in 16,385 flying hours. The first squadron to receive the Hastings was 47 Squadron, and was also the first Hastings squadron on the Berlin Air Lift.

The first production aircraft had been delivered to the RAF less than twelve months after the prototype had flown, 47 Squadron's first aircraft arriving at Dishforth in October 1948. With a top speed of 354 mph the Hastings was faster than the heavy bombers then in service with RAF Bomber Command. On 1 November 1948 eight of 47 Squadron's Hastings flew from Dishforth to Schleswigland to join the Berlin Air Lift. Most of the pilots had been rushed through a conversion course, some having only 30 hours flying time solo on type. In one instance a pilot flew 365 hours on the Hastings in six months after his first sight of a Hastings - he was also the first Hastings' pilot to complete 100 sorties to Berlin. At one time there was such a shortage of pilots with Hastings experience that volunteer Glider Pilot Regiment pilots were flying as second pilots.

An average of twenty-six Hastings were available for the Berlin Air Lift operation, but the number based at Schleswigland never exceeded twenty-four, and was usually well below that number. Unfortunately the effectiveness of the aircraft was reduced, not by unserviceability but by the shortage of trained aircrew and the RAF's policy of returning the aircraft to the U.K for every 50 hour inspection. In regard to unserviceability, the number of defects per flight decreased rapidly as more time and experience was gained, and likewise the repair time per defect showed a considerable reduction from the beginning of April 1949 onwards. Crews in general liked the Hastings and the groundcrews appreciated the accessible servicing points, but what was not appreciated was the lack of spares and the bad cockpit lighting. The payload was initially some 16,000lbs, which was governed by the permissable landing weight of 70,000 lbs, but later on the permissable landing weight was increased to 74,000 lbs at a take-off weight of 78,000 lbs, and so the payload was increased to between 19,500-20,000 lbs.

A few interesting facts of this period of operation indicates how in its initial period of service the Hastings, one of the last 'tail-draggers', was to prove a heavy hauler. The Hastings flew a round trip of 398 nautical miles (Schleswigland to Berlin - and return) in 2 hours 36 minutes carrying a load of 9.3 to 9.4 short tons, whilst the USAF DC4 Skymasters flew a round trip of 276 nautical miles (Fassberg to Berlin and return) in 1 hour 50 minutes carrying 9.9 short tons. With regard to the Hercules 101/106 engines, these were just introduced prior to the operation, yet had a lower failure rate of 0.11 per 1,000 flying hours than the Merlin T24 engines on the Avro Yorks, whose failure rate was 0.96 per 1,000 flying hours. The larger proportion of flying time spent at high rpm for take-off and climb, as opposed to

Three Hastings C.Mk.1's in formation after acceptance into the RAF [J. Knivett]

HANDLEY PAGE HASTINGS AND HERMES

normal Transport Command flying, appeared to affect the Merlin more than the Hercules.

In January 1949 petrol was getting short in Berlin and it became necessary to lay on eighteen Hastings flights in a hurry whilst civilian aircraft were specially modified for the carriage of liquid fuel. Each Hastings carried twenty 40 gallon drums on these sorties. A further out-of-the-ordinary load for the Hastings were numerous girders and rolls of cooling tubes for a West Berlin power station. Some of the girders were 1 1/2 feet wide, 4 feet deep, 32 feet long and weighed nearly 7,000 lbs. These were loaded by crane into the entrance of the Hastings, from where they were man-handled into the fuselage interior.

297 Squadron were the next unit to commence equipping with the Hastings and joined the Air Lift in August 1949, followed shortly afterwards by a detachment from 53 Squadron, who had also commenced equipping with the Hastings. The next unit to receive Hastings aircraft was 99 Squadron at Lyneham, who commenced receiving their Hastings in August 1949, followed by 511 squadron in the September, also being based at Lyneham.

The practice of allocation of aircraft to bases instead of squadrons commenced during the Berlin Air Lift and became standard practice afterwards. By the end of the Air Lift the Hastings had carried over 55,000 tons of cargo, large quantities of which was coal, which must have been frustrating to the crews - to say the least - to see their nice new Hastings loaded with bags of coal. The main damage to the interior structure concerned some lashing points and under-floor structure, but nothing that made the aircraft unserviceable.

Although the Hastings could and did take-off in gales, the main difficulty was in taxying in crosswinds, for without the control locks 'in' the gusts could force the rudder to its extreme range against the pilot's foot load. The control locks were spring loaded to the disengaged position, so that upon moving the locking lever in the cockpit to the 'off' position, the locks would disengage and the control surfaces free to move under the pilot's control. The controls would then be moved to check their freedom of movement to ensure that the locking plunger had disengaged - this will become more apparent later when discussing a number of accidents where the locking plunger had not disengaged. Taxying across wind with no locks on the controls could become quite hazardous, as the large rudder under the influence of the wind could almost kick a pilot out of his seat! Further damage caused by the high winds experienced at Schieswigland was to the control surface spring tabs, for even with the control locks 'in' the spring tabs were still free to move on their torsion bars.

The first accident to a Hastings was achieved by 47 Squadron on 2 October 1948 at Dishforth. TG519 whilst being flown on a three-engined practice with number 4 engine shut down and the propeller feathered, encountered extremely turbulent wind on the approach, causing the aircraft to undershoot with the port wing striking some trees, then the aircraft hit the runway too hard and the undercarriage collapsed, the aircraft next striking the runway control van before coming to a halt.

The only serious accident to a Hastings on the Air Lift, in which all on board were killed, occurred when TG611 took-off from Berlin's Tegel airfield with 'nose up trim' still set. This resulted in a steep climb to the stall and a nose-in crash. Apart from this a number of other accidents and incidents occurred, in one case TG510 was forced to land with its wheels up after an incident on take-off. It was officially considered that the aircraft had touched down again on take-off as the undercarriage was retracting, this damaged the retracting mechanism units and made retraction or extension impossible; but the aircraft was repairable. Failure of the tailwheel fork locking grub-screws also resulted in a number of incidents, this was suspected to be due to the towing crews ignoring the locking device when tail-towing the aircraft. In one case this resulted in TG530 crashlanding due to non-retraction of the tailwheel.

An avoidable accident on 6 April 1949 rendered TG534 to almost a burnt-out shell at Schleswigland, fortunately without loss of life. During servicing of the aircraft a ground fuel cock was accidently left open, so that when the main fuel cocks were turned 'on' upon starting the engines, fuel poured out from the open cock and was ignited by the engine exhaust flames. The fire spread through the wing roots and could not be

Hastings C.Mk.1's of Transport Command parked on the Schleswigland tarmac in 1948 [J. Knivett]

SERVICE USE OF THE HASTINGS

Hastings TG510 following its crash-landing after a take-off incident as described in the main text. [J. Knivett]

stopped, only being prevented from reaching the aircraft fuel tanks by prompt and courageous action of the fire service.

During the Berlin Air Lift the Hercules 101 engines were a credit to the Bristol Company, operating opproximately 40,000 hours with only one engine failure. Likewise with the hydraulic system and undercarriage, these were quite satisfactory with not one oleo strut being changed, although leaking jack glands did give a little trouble. The airframe weatherproofing was very bad with leaks into the cabin, and electrical equipment was a source of trouble. With respect to the flap electrical and undercarriage indicator warning light systems, both required constant need of maintenance due to faulty micro-switches. One instrument system that was in need of urgent modification was the engine oil pressure indicating system, this proved most unreliable, and the Handley Page representative at Schleswigland, A.G 'Jack' Knivett (known to the RAF crews as 'Sir Fred') had to deal with approximately ninety cases. This problem alone was the frequent cause of aircraft delays and the expenditure of many man-hours. Another factor that did not help maintenance was that the groundcrew as well as the aircrew were new to the aircraft, and every three months were replaced with a new set of men without Hastings experience.

The airfield at Tegel in the French sector of Berlin was built on rubble from the city ruins and the ground alongside the runway was soft, this combination could prove hazardous. In one case a

*Above and left:
Steel girders for s Berlin Power house are prepared for air shipment aboard an RAF Hastings. Each of the 7,000 pound pre-fabricated structure had to be manoeuvred through the Hastings freight door. The railway waggons in the left photograph were used to bring cargo directly onto the airfield.
{J. Knivett]*

HANDLEY PAGE HASTINGS AND HERMES

The result of a take-off incident at Tegel due to failure of the tailwheel self-centering device. The aircraft swung off the runway and onto soft ground. [J. Knivett]

Hastings had a grub-screw fail in the tailwheel assembly, preventing self-centring of the tailwheel. This proved almost disastrous, for the failure caused the aircraft to swing off the runway to a point approximately sixty yards away into soft ground, digging in the port mainwheel and bending a propeller blade. The USAF who operated the airfield on behalf of the French, gave the Handley-Page representative six hours to move the aircraft, failure to do so would result in U.S engineers blowing it up. With the aid of two RAF personnel and German labour, and against argument from the U.S control staff, the effort succeeded after 8 1/4 hours. A trench had been dug down to the port wheel, the bent propeller blade straightened by brute force, and the aircraft pulled out. On take-off there was vibration from the engine, so the propeller was feathered fifty yards on the take-off run and the aircraft returned safely to base for inspection.

In one single month twelve Hastings flew into Berlin nearly 3,000 tons of coal, 1,400 tons of mixed freight and nearly 160 tons of powerhouse equipment. 47 Squadron over the Lift period for instance carried 22,000 tons in 3,000 flights. One Hastings navigator summed the aircraft up thus: *"The Hastings is well suited for Lift duty and from the navigator's point of view is a wonderful aircraft. It has been designed for transport work and that means a comfortable aircraft. Equipment and crew stations are very conveniently located. The main thing we all like about it is the speed range; one can fly between 130 and 240 knots fully loaded. Thus if your block* (*Aircraft flew in blocks, each block a specific aircraft type, to prevent late or early arrivals.) is late starting you can still reach your beacon on time. Our block was once 25 minutes late and we still made the beacon on schedule. Not every navigator can make a claim like that for his aircraft".*

The remains of Hastings TG534 at Schleswigland after a start-up fire caused by the failure to lock the fuel drain cock. [J. Knivett]

SERVICE USE OF THE HASTINGS

Three Hastings on a low-level flypast over Schleswigland on completion of the Berlin Airlift in 1949. [A. Mitchell]

24 Squadron commenced re-equipment with the Hastings in December 1950 and was based at Lyneham. Then came a re-organisation of the Hastings bases, one at Lyneham with 53, 99 and 511 Squadrons and the other at Abingdon with 24 and 47 Squadrons. Lyneham was listed as the strategic support base and Abingdon the tactical base, with 24 Squadron still being responsible for VIP flights. Crews were converted on to type by the Transport Command Operational Training Unit (TCOTU), which was 242 OTU based at Dishforth; the standard course for conversion taking sixteen weeks. The course usually finished with an overseas flight to test individual crew members, and then the individual members were posted to their Hastings squadron to join their crews.

In Coastal Command at this date 202 Squadron were still flying their Halifax Met' aircraft from Aldergrove in Northern Ireland, carrying out weather forecast flights over the North Atlantic. Then in October 1950 they began to equip with Hastings Met I aircraft and became fully re-equipped in the November. These meteorological sorties were code-named 'Bismuths', and were of nine hour duration, sometimes covering up to 3,000 miles in one sweep. The Hastings for these duties were specially equipped to observe and record weather, taking temperature and pressure readings, with camera recordings of cloud formations. The patrols had a double function also, for each aircraft carried air-sea rescue equipment and

Hastings Met Mk. 1 TG616 on a test flight before delivery. [J. Knivett]

HANDLEY PAGE HASTINGS AND HERMES

emergency supplies and articles.

The Hastings for the met' duties had the second pilot's position converted for use as a met' observer's position also, with each aircraft carrying a qualified met' observer as on the Halifax but this time with a bit more comfort! The remainder of the crew had almost airline standards of comfort as well, and although the sorties were of an exhausting nature, with constantly changing altitudes of anything from 200 to 18,000 feet, the capacious cabin allowed the crews to stretch out, relax on a bunk, as well as prepare hot food and drinks. The weather information was radioed back to base for onward transmission, and the data supplied used to assist in weather forecasting. Sixteen aircraft were converted to Met I standard and were initially painted grey overall, but this was later relieved with a white cheat-line down each side of the fuselage. Some were painted grey with a white underside. Typical aircraft on this duty was TG572, TG620, TG621 and TG624.

By the end of 1950 two landing accidents had occurred with the Hastings, one of these being TG583 of 241 OCU, which crashed on the approach to Dishforth on 31 July. The other aircraft was TG574, which crashed on the approach to Benina airfield in North Africa. Both aircraft were destroyed, but only in the latter case were there fatalities. During the first year or so of operating the Hastings, a number of engine shut-downs and precautionary landings were carried out, which led to the aircraft being nicknamed *"The greatest three-engined transport aircraft in the world"* - a name transferred from the early operating days of the York. The Hastings also received a compliment from the Press, one newspaper report stating :

"There is nothing of the carthorse about the Hastings, in fact, it has the grace and beauty of an airliner and behaves like one".

Regarding a number of the engine shut-down incidents, some were due to faulty instrument indications, some to new crew members unfamiliar with the aircraft, whilst others were due to loss of engine oil or high engine temperatures. These latter cases were quite a surprising factor seeing as the Hercules Mk.100 engines on the Halifax had by then achieved a high reliability reputation for piston engines.

In regard to the engine troubles, these centred around two problems, one known as 'gulping', which was the loss of engine oil through the breathers. The other was 'coring', which was when engine oil passing through the oil cooler at too low a rate of flow or too low a temperature started to congeal in the oil cooler matrix. Provided that a 20 mesh filter was fitted to the engine, 'gulping' could be caused by an indiscriminate mixing of additive and non-additive oils, or by the formation of water in the engine oil system caused by condensation of atmospheric moisture and the products of combustion within the engine. When the water reached boiling point it vapourised and prevented normal scavenge pump operation, the steam dislodged the carbon deposits which clogged the filters, causing the oil to build up in the crankcase and become discharged through the breathers. Prevention of this was by maintaining the engine oil temperature at or above 70 degrees C during cruise conditions. Mixing of the additive and non-additive oils caused the combustion products such as carbon to be disturbed and block the filters, with the same results.

"Coring" is the condition that occurs in the oil cooler when the flow through the matrix is slow and outside air temperatures are low. The oil then congeals in the matrix tubes and prevents the passage of oil. This is prevented as with 'gulping', and if 'coring' does begin, then the rpm is increased and the boost lowered, this raises the oil temperature and flow rate, so reducing the possibility of complete congealing.

Returning once more to the routine flying of the Hastings, a typical operation was the transportation of an RAF photographic survey team with all its personnel and equipment from the Gold Coast in West Africa to Kenya, a journey of 3,500 miles. The journey to West Africa being used to ferry troops and supplies to that region. Included under typical operations, although unscheduled, was the transportation of VIPs, depending on who was being transported, for these could sometimes be formal tours. With the introduction of the C.Mk2 would be introduced special interior furnishings, which in the end led to the C4 VIP version of the Hastings.

Air Support and communications.
The bulk of the Hastings C Mk.1s were in service by 1952, and with the introduction of the C Mk.2s the aircrew found that the aircraft handling had improved and the Hercules 216 engines eliminated most of the engine problems. The Hastings has been called "The greatest leveller of all aircraft", for pilots who were absolutely brilliant on other aircraft sometimes failed to make the grade on the Hastings. Its swing on opening-up for take-off was more vicious than either the Halifax or Lancaster, and no doubt upset the ego of many a pilot, but it could be held - the swing not the ego! Yet it has been classed as a good, honest, rugged transport aircraft by most who flew her, and she proved to be a steady platform for instrument flying. One pilot said of the Hastings:

"It was built like a brick loo, it was as steady as a brick loo, but I was cautious when landing and taxying - she could catch you out if you relaxed".

It was designed for paratroop dropping and could carry the Heavy Stores Beam with an Army gun and Jeep slung from it and supply containers outside of it. In this configuration Transport Command aircrews muttered of "built-in headwinds", yet the same load had been carried by 38 Group Halifaxes on operations during the 1939/45 war, but very little use was made of this configuration on the Hastings, the Suez War operation possibly being the exception.

The RAF received its last C Mk.1s during October 1952, and these continued on operations

SERVICE USE OF THE HASTINGS

A 5 cwt 'Jeep' being loaded onto the Heavy Stores Beam on Hastings [Author]

on numerous trooping and evacuation sorties over the years, including the casualty evacuation of Korean War wounded by the Far East Air Forces. These British wounded were shipped to Iwakuni in Japan and then flown to Changi in Singapore, where they joined the Hastings service of the Far East Air Force's casualty evacuation scheme. Each Hastings normally carried twenty casualties with a flight sister and nursing orderly. These Hastings were fitted out as flying hospitals, and normally flew twice a month to the U.K, the route being from Singapore via Negombo, Habbaniya, Fayid, Luqa to Lyneham. Casualties for the service also came from Hong-Kong, North Malaya and Ceylon. The same aircraft flew troops from the U.K to the Far East on their outbound journey, as well as stores of an urgent nature for the bases en-route.

The fighting in Malaya against Chinese Communist terrorists also called for the use of the Hastings for supply dropping to troop columns. This involved flying at low altitude to drop 'free' packages; dropping the packages at heights of 300 feet in areas with trees of 100 feet height was no mean feat, as some of the packages weighed up to 500 lbs and were bulky.

Prior to this, on 20 December 1950, an accident occurred which should have erased any doubts about the longitudinal stability of the Hastings. TG574 took-off from El Adem airfield shortly before 2000 hours piloted by Flight Lieutenant Tunnadine, who set course for Castle Benito, his next stop. The aircraft and crew, plus a slip crew, were returning from a route proving flight and had staged through from Singapore. Tunnadine had been rated as above average ability, and before the night was out he was to prove exactly that and more. At 8,500 feet the captain called for the flight engineer to give him cruising power, then sent his second pilot back to the rest bay and asked Squadron Leader James, a fully qualified Hastings captain, to come forward.

At 2045 hours there was a loud bang in the crew compartment, an acrid smell of burning, a severe shudder through the aircraft. The next thing was that Tunnadine found that he had no rudder, elevator or trim controls, just lateral control. Due to metal fatigue one blade of the port inner engine's propeller became detached, and out of a possible 360 degree exit had sliced into the fuselage; severed all the controls to the tail control surfaces, and severely injured Flight Lieutenant Bennett, the second pilot, who was resting. The loss of the blade set up out-of-balance forces which tore the port inner engine from its mounting, and it fell away.

The situation now took on the makings of a disaster, only three engines functioning, only lateral control, and drag from a nacelle without an engine. The aircraft started to turn to port and lose height, Tunnadine called the quartermaster to move some of the passengers to the rear and stand by for further orders. Next the baggage was moved to the rear, and slowly the nose started to come up again. By judicious movement of baggage

HANDLEY PAGE HASTINGS AND HERMES

An Army RAF casualty evacuation flight with a person in an 'Iron Lung' to assist breathing is disembarked from Hastings TG529. [P.Porter]

and passengers the aircraft was at last trimmed for level flight.

An SOS had been sent out and acknowledged by Benina, and very, very gradually Tunnadine manoeuvred TG574 in the direction of Benina, but the odds were very heavily stacked against him, probably more than he knew. It was now dark, and the decision now had to be made where to land, for neither crew nor passengers had parachutes. Over the next few moments information on the locality and obstructions passed from Benina to the aircraft. A medical officer on board the aircraft, Squadron Leader Brown, had meanwhile moved forward to try and extricate Flight Lieutenant Bennett from the wreckage in the restbay, but this was found impossible. Then Tunnadine ordered Brown back to his seat for a possible crash landing, but he requested and received permission to remain where he was with Bennett.

Gradually Tunnadine started to lose height on his approach to Benina, but was not able to use the flaps because of the change of trim they would incur. Escape hatches were removed and the decision made to land on the runway at Benina. At Benina the rescue services had been alerted, the number of crew and passengers ascertained, a request made to Benghazi for extra medical staff, which was quickly actioned. The flarepath was lighted and the rescue service vehicles stood by near the end of the runway with engines running and lights extinguished. Tunnadine was now down to 1,000 feet and on a reverse bearing from Benina and preparing to make a 180 degree turn to line up with the Benina runway. The strain on Tunnadine must have by then exceeded anything that a pilot was asked to bear, yet still he bore it, calmly answering the R/t.

At last Tunnadine had the aircraft lined up with the flarepath, crew and passengers tightened their safety belts. The three engines had by then exceeded their maximum permissable temperatures, nearer and nearer to the ground came the aircraft, still in a level attitude. Then came Tunnadine's last R/t call that he could not see the end of the runway. TG574 hit the ground 600 yards short of the runway, debris from the engine nacelles and the underside of the fuselage flew into the air, for in spite of Tunnadine's last application of power there was no response from the engines to lift the nose. For about 100 yards the aircraft tore across the ground, then became airborne again, a wing dropped and was torn off, the aircraft rolled on its back and slewed around.

Immediately the rescue services swung into action, yet some of the survivors were already emerging, foam was poured over the engine areas, crash-crews tore into the fuselage to search for more survivors. Only the flight engineer of the operating crew survived, Squadron leader Brown was seriously injured but recovered and continued in the Service, none of the passengers or the quartermaster were injured. So the fine airmanship and the professionalism of his crew

SERVICE USE OF THE HASTINGS

resulted in the survival of 36 passengers and two crew members.

In the aftermath of the accident a number of points came to light. Tunnadine had been notified of the undershoot area and that it was not lighted, but the flarepath party had thought they would help by putting flare-lamps on the undershoot area, and the QNH given to the pilot was incorrect due to a faulty instrument. The approach to the runway was slightly rising ground with a rocky outcrop - the odds were already heavily stacked. Whether these other factors would have had any affect on what happened is open to conjecture. Survivors praised both Tunnadine and the strong structure of the Hastings. One of the slip crew members was R/O Ken Heseltine, who felt that the Hastings was one of the safest aircraft, continued flying on the Hastings to complete his full service in the RAF.

In 1952 two Hastings were detailed to depart to Thule in Greenland, these were WD490 of 24 Squadron captained by Flight Lieutenant D.Waight and WD492 of 47 Squadron captained by Flight Lieutenant M.Clancy. Their task was to supply the British North Greenland Expedition at Northice Camp. Having established themselves at Thule, the two aircraft began a number of supply drops.

On the 16 September Clancy and crew on WD492 with eleven personnel on board took off for Northice. Twelve paradrops were successfuly made and then the free drops commenced, these were from 50 feet. Dropping was to be made on the aircraft's radio altimeter and with Pilot Officer Taylor on the ground radio monitoring the drops and giving visual observations of height.

On the second run-in Clancy started to let down, reduced his speed to 125-130 mph, the packages went out and WD492 started to climb away - suddenly a whitish void, 'white-out' - the port wing touched the ground and dug into the snow. Clancy tried to lift the wing, using so much force he bent the control column. Then the aircraft smashed onto its belly, skidding along in great big groundloops for approximately 12 miles. Both port engines were snapped off, the flight engineer pressed the fire extinguisher buttons as the aircraft came to a halt, the fuselage completely intact, but a bent-up port outer wing. More than one crew member was reported to have blessed Handley Page for making the Hastings a strong aircraft - in between breathing a sigh of relief! The Northice team hurried to help, but fortunately only three of the Hastings' complement were injured and were settled inside the fuselage to keep warm.

Rescue was to be a difficult task, and certainly could not be carried out by the RAF. After seven days a USAF Rescue Albatross amphibian skidded in on its keel, and after a lot of hard work preparing the area for take-off, the Albatross was loaded and a risky take-off accomplished with the three injured personnel. Two days later a USAF skid-equipped Dakota arrived at Northice and carried out a very risky landing in bad conditions, fully in accordance with the traditions of the USAF Rescue Service. JATO units were then bolted on to the Dakota's fuselage, and with full power and the JATO units blasting away the remaining Hastings personnel were lifted out.

As well as all of the personnel surviving - and leaving a Hastings up in Greenland for posterity, with possible retrieval by keen enthusiasts in the years to come - the RAF Transport Command continued with their re-supply of the Greenland Expedition at Northice and flew in 86 1/2 tons of supplies.

During the service of the Hastings the new aircrew role of loadmaster was introduced, this combined the work of steward when carrying passengers and the ascertaining of the correct and secure loading of baggage and cargo, the name being changed to quartermaster later. It would appear that during the early service of the quartermasters their enthusiasm outweighed their knowledge of the job, but after gaining experience they were considered an asset to the crew complement; so the following two stories from the aircrew must be taken against the latter section of the statement.

Following airline practice it was normal after take-off for the quartermaster to arrive in the crew compartment with cups of tea for the operating crew. In one case an over-enthusiastic and young quartermaster arrived in the cockpit with a tray of cups of tea. Seeing no place to rest the tray he raised some levers to the horizontal position - as these were the engine fuel cut-off cocks there was a mad scramble to get the levers to the 'on' position before there was an ominous silence from the engines - the cups, tea and tray finished on the floor amidst rude remarks.

In another instance, after a rather hectic and bumpy period in the flight the quartermaster appeared with refreshment, the captain noticed that the quartermaster's head was cut and bleeding. When told that this had occurred during the bumpiness and he had hit his head against the roof racks, the captain told him to go and bandage it. Shortly afterwards, the captain going to the rear found the rack well bandaged and the quartermaster still cut and bleeding! This is not to decry the Hastings' quartermasters, but to illustrate their enthusiasm for the job and integration with the crew. This did occur and they proved their worth, so well illustrated in the case of the tragic crash at Benina.

Flight engineers were also on occasions to receive 'rockets' from the aircraft captains, one flight engineer who wishes to remain nameless remembers a flight with a rather high-ranking pilot. Having watched the fuel 'burn up' during the flight and having warned the pilot of their endurance left, he was told that there was plenty of fuel for the flight. However, with the weather not on the bright side and the fuel getting low the flight engineer persisted with his warnings. In the end the pilot was forced to divert to another airfield - the flight engineer was then given a 'rocket' from the pilot for not having warned him in time! Which goes to prove that you cannot win

HANDLEY PAGE HASTINGS AND HERMES

all the time, infact, the flight engineer was lucky to win any time.

During 1951 a turn-about took place in North African affairs when the Anglo-Egyptian Treat was abrogated, inside the month 10,000 soldiers, 350 vehicles and guns and many tons of supplies were flown into the Canal Zone. During the second month a small force of Hastings carried 21,500 troops and other passengers within the Middle East as well as transporting 725,000 lbs of cargo. During the most intensive period of flying the Hastings were each flying 18 hours a day, and the largest collection of RAF transport aircraft ever gathered were seen at Fayid airfield (Canal Zone). Fayid would also be the gathering place for Hastings aircraft during the Abadan crisis, which was the first foreign nationalisation of British interests.

The Hastings was in the 1950s to prove the backbone of RAF Transport Command and provide the means of mobility for quick reinforcements of men and material. In 1953 the Hastings was the means of flying in troops and supplies to counter the Mau-Mau emergency in Kenya. In the February the scene had changed for a home emergency, for with the disastrous floods in East Anglia, the squadrons were engaged in flying in hundreds of tons of sandbags.

During landing at Malta on 16 June 1952, TG603 was blown-off the runway, crashed, and was damaged beyond repair. Then on the 12 January 1953 came the first structural failure on an Hastings. TG602 of the Airborne Support Flight at Abingdon was on detachment at Shallufa airfield. It was involved in carrying out steep turns at about 2,000 feet about the airfield when eye-witnesses saw parts of the tailplane and elevator detach themselves from the aircraft. The Court of Inquiry into the accident determined that the cause originated with the detachment of the starboard elevator, following which came the break-away of the port elevator and tailplane under the induced stresses and strains. The aircraft had then gone into a dive straight into the ground, killing all on board. Also ascertained during the inquiry was the fact that modification 801*(Mod 801 replaced the 1/4 inch diameter bolts with 5/16 inch bolts.) had not been incorporated, which had been brought in to prevent this type of structural defect.

On the subject of structures and systems, it might be of interest to mention the fuel system and different re-actions of flight engineer's. The Hastings fuel system, as explained in Chapter 5, was very similar to the Halifax system, and though understood by ex-Halifax flight enginers was not appreciated by ex-Lancaster flight engineers, who were more used to the - rather basic fuel system of that aircraft. With the multiple fuel tank system of the Hastings/Halifax, extra work was entailed in 'changing over' tanks and working to a fuel management system. This was more than compensated for by knowing that the loss of one tank through malfunction or enemy action, only reduced the fuel capacity by that amount (one of fourteen tanks), while the loss of one fuel tank on a Lancaster made a vast difference in capacity (one tank out of six). It can also be mentioned that

Transport and Coastal Command Hastings parade at the Queen Elizabeth II Coronation Review at Odiham in 1953.

SERVICE USE OF THE HASTINGS

no Hastings was lost through 'pulling' the wrong fuel lever or fuel mismanagement. The fuel management contributed to main plane spar life by providing a wing relieving load and thus reducing fatigue. To contribute to this the outer fuel tank groups were always filled to capacity where possible.

Two Hastings joined the RAF College at Manby in 1951 and 1952, WJ327 and WD499 respectively. These were involved in a wide training programme over a number of years, including navigational and reconnaissance flights, investigation and development flights, as well as liaison and experience exercises to other countries. Flights being made to both the tropics and artic regions. On one flight WJ327 returned from Accra in West Africa direct to Manby, a distance of 3,300 miles in 15 hours 20 minutes. This was followed in September 1952 with the commencement of specialist navigational exercises to RCAF Resolute Bay and USAF Thule. From these airfields the Hastings made flights over the North Pole and so gained experience in maintenance and navigation in extreme cold, as well as the effect that the Magnetic Pole had. On early flights there, inspite of pre-heating and oil dilution*(* Oil dilution was the dilution by petrol for easier starting.), three engines were wrecked during ground running; which determined the need to extend the oil dilution times and to increase the ground-running period of up to forty minutes.

Sometimes unfamiliar routes were explored, such an example occurred in March 1953, when WJ327 carrying instruction staff and pupils flew to Korea. On the return flight the decision was made to survey a route aross the Indian Ocean to the airfield at Johannesburg in South Africa. QANTAS Airways had at that time commenced a pioneering service through the Cocos Island and Mauritius, but it was not at that time a military route. So WJ327 was flown from Hong-Kong to Cocos in one hop at night, crossed the equator at 25,000 feet in a storm and emerged at Cocos dead on track - which was just as well, as the next land would have been Antartic! The next course was to Mauritius and a night flight of 2,700 miles of ocean, without the aid of radar or radio beacons. With the use of basic navigation and astro shots Mauritius came up dead ahead through the dawn, with a flight time to Mauritius of 114 hours. It was even more gratifying to find out that their flight time was only 1/2 hour longer than that by the new Super Constellations of QANTAS.

Even with the addition of the Canberra to the College's fleet the Hastings stayed in attendance on flights overseas, until finally, in June 1955, the two Hastings performed their last duty with the College. This was in the support role to the long range *Aries IV* Canberra in its first trans-polar flight by a jet aircraft. WJ327 covered the departure point at Bodo and WJ499 positioned itself off the Alaskan coast, their role to pass on weather information and to maintain radio contact as necessary. The flight was a success, the Hastings did their job and moved back to normal transport operations at the end of the month.

On 22 June 1953 WJ335 had taken off from Abingdon and was the first of a number of incidents and accidents that related to the elevator lock. It became airborne after an unusually short run, climbed to about 300 feet where it reached the stalling point, and with engines at full power dived into the ground. The court of inquiry after the accident investigation, placed the cause as the failure of the pilot to release the elevator lock (which positioned the elevator approximately 10 degrees in the 'up' position) and failed to check that they were removed. In spite of this inquiry and notification of modification 875, introduced to prevent such an accident, further trouble did occur. One of the last was to WD484 at Boscombe Down on the 29 March 1955, the aircraft had been detached to RRE on radar trials. The events followed a sequence similar to WJ335, except that at the stall the port wing dropped and the aircraft went into a violent roll to starboard, then dived into the ground and broke in two. Two of the crew were killed and two seriously injured.

With a series of engine failures related to 'gulping' on the Far East route, it was decided to carry out an extensive investigation. With this in view a group of RAF engineer officers were flown in TG613 to carry out on the spot investigations. On the 22 July 1953 whilst flying between Idris and Habbaniya TG613 had three engines fail - all suspected to be due to oil gulping! One could say first-hand experience, as well as becoming the only case of ditching of a Hastings! The ditching was successful and all the crew and passengers survived. Bristols had in anycase already investigated this problem as a number of other Hercules powered aircraft had suffered 'gulping' incidents.

Within five days a further Hastings was written off, this was TG564 and the accident was at Hong-Kong Kai Tak airport. The pilot undershot the runway and then proceeded to carry out a demolition job on a number of obstructions, finally crashing through the boundary fence and catching fire. Due to inadequate fire equipment on the airfield and incorrect crash procedure by the flight engineer, the aircraft was severely damaged and beyond repair.

This accident was followed by a further incident on the 17 July 1953. WD489 took off from 5 MU High Ercall for the short flight to RAF Kemble. On the approach to Kemble it was found that on the runway in use there was a 60 degree crosswind from port of 20-25 knots. The approach was made at 130 knots, reducing to 100 knots over the boundary. No difficulty was experienced in correcting for drift and full flap was selected at 500 feet. Round out was carried out to make a taildown landing, but the pilot checked rather high and the aircraft settled heavily but did not bounce. On back-tracking down the runway the starboard tyre deflated, and only on disembarking did the crew see the wrinkles on the starboard mainplane. Further investigation found that the

HANDLEY PAGE HASTINGS AND HERMES

Hasting C. Mk.2 WJ334 of 36 Squadron Transport Command [J. Knivett]

starboard undercarriage shock absorbers had bottomed and the starboard undercarriage attachment had buckled. The total hours flown by the aircraft were just over 17 hours since construction, but it is nice to report that WD489 continued in service after repair with 24, 47 and 70 Squadrons until struck off charge in 1968.

Only one accident occurred with a Hastings in 1954 and that was due to pilot error, this was to WD482 and did not destroy the aircraft. In the meantime the Hastings continued operating all over Transport Command routes, transporting troops to the Commonwealth's 'hot spots', quietly performing their tasks over the Atlantic and performing airborne support duties where-ever they were required.

During 1955 five Hastings of 511 Squadron played nursemaid and rendered support to 139 (Jamaica) Squadron, when that unit's Canberras participated in a goodwill tour of the Caribbean area and took part in Jamaica's celebrations. The tour was led by Air Vice Marshal J.Whitley, AOC 1 Group Bomber Command, who at the end of the tour complimented Transport Command Hastings for their regularity and dependability. The Hastings were operated into a number of limited airfields, as well as to Georgetown in British Guiana, giving flying displays over and at a number of towns.

During the flight to Miami outbound, and on the flight to Nassau, brief encounters were made with two hurricanes, causing the flight to arrive one day late at Bermuda. Clear weather at Bermuda made it possible to give a successful flying display over Hamilton. The return flight via Canada was quite straight forward and while the Hastings were at Montreal they ferried loads from Manitoba, Winnipeg and Goose Bay, Labrador. The Hastings force leaving for the U.K flew through Goose Bay, with two aircraft flying via Narsassuak in Greenland and the others flying via Keflavik in Iceland. During the tour the Hastings had flown 1,420,000 passenger miles and carried 172 gross tons while flying 77,000 miles.

In early 1956, units operating the Hastings were 24, 47, 53, 70, 99, 202 and 511 Squadrons, then in March the Beverley entered service and began to equip 47 Squadron, the relinquished Hastings were then available for delivery to re-equip 70 Squadron in Cyprus, who had been operating Valettas. 1956 also saw the confrontation between Britain and France on one hand and Nasser of Egypt on the other, civilians being evacuated from Egypt by British civilian airliners whilst diplomatic moves went ahead. By October there were stationed in Cyprus 70 and 99 Hastings Squadrons, soon to be joined by 511 Squadron.

The Suez War campaign, named 'Operation Musketeer', began on 31 October with the destruction of the Egyptian Air Force, military targets and troops. Then on the 5 November Hastings loaded with paratroops and supply containers took off at 0500 hours at 20 second intervals from Cyprus, some of the Hastings carrying gun and jeep loads slung underneath. The British drop was made at El Gamil airfield, two miles west of Port Said, the troops being the 3rd Parachute Battalion commanded by Lt Colonel Paul Crook, being carried by sixteen Valettas and fourteen Hastings. The drop commenced at 0715 hours and was over in 125 minutes, the Egyptians fleeing east to Port Said.

Amongst the equipment dropped by the Hastings were six 106mm recoilless anti-tank guns, 3 inch mortars, seven Jeeps, four trailers and 176 supply containers. The paratroops were in possession of the airfield inside 30 minutes, yet no attempt was made to air-land any other troops. A solitary Hastings later flew over Cairo and dropped half-million leaflets urging the population to accept Franco-British proposals. Hastings were also used to ferry casualties back to hospitals in Great Britain.

SERVICE USE OF THE HASTINGS

Hastings C.Mk.1A TG75 of 70 Squadron over Cyprus in 1962. [B. Gardner]

During 1956 a Hastings was also to be involved in an unusual and embarrassing incident. This occurred on 4 July when a C4, WD500 flown by Flight Lieutenant Graham and carrying Sir Gerald Templar and other VIPs eastwards, suddenly had its passenger door complete with hydraulic steps break away in flight. The steps separated from the door and fell clear, but the door hit the port tailplane and remained in position against and across the tailplane leading edge. The result of this was a severe porpoising effect that required full fore and aft movement on the control column to maintain control - the flight engineer monitoring the effect estimated that the undulations were of the order of 1500 feet. The pilot reduced speed and power and in so doing reduced the porpoising to a manageable quantity, sufficient to allow him to make a normal landing at the nearest airfield, where the door fell off the tailplane during the landing run.

The C2 and C4 aircraft were in general preferred by most crews as they were more longitudinally stable, although development had improved the earlier longitudinal instability and the engine problems of the C1. A.N.Mitchell, an ex-Halifax, Hastings and Britannia flight engineer, said of the Hastings:

"The Hastings was alright, although the C1 could be a pig, for we originally suffered with coring on the engines and carburettor icing, but it was not a bad aircraft at all, it was structurally sound and turned out a reliable aircraft". This was borne out by other crew members, some who later flew on the Comet and found that aircraft not as comfortable as the Hastings. Most liked the Hastings, which once over its initial problems settled down to a good safety record that was only marred by a bad accident in 1965.

The passenger seats facing aft may have seemed strange on the ground with the tail down, but was one of the best safety factors during any accident. Squadron Leader 'Jacko' Jackson talking of the Hastings:

"It was known in its time as "The greatest leveller of all aircraft". This aircraft was probably the last of the six crew configuration, which comprised of captain, co-pilot, wireless operator, navigator, flight engineer and loadmaster. The captain's and co-pilot's seats were the most luxurious ever built, moving forward, upwards, downwards, armrests, headrests, you name it. When designers asked the RAF what seats to fit, they were told "Fit Hastings' type seats"".

Another crew member of Hastings was radio operator Ken Heseltine, who after discussing the TG574 crash at Benina and his escape, said of the Hastings:

"I had faith in the Hastings' construction, and still consider it one of the safest aircraft, what happened was just one of those things. It turned out a reliable aircraft and I was happy to fly on them".

1957 commenced with the proposed introduction of Britannia aircraft into RAF service, with which the Hastings base was transferred from Lyneham to Colerne; 24 Squadron moving there in the January and still operating VIP flights, with 511 Squadron moving there shortly afterwards. With no new confrontations over the year, 1957 was to be a relatively quiet one for the Hastings, to allow them to prosaically continue flying the trunk routes conveying material and personnel. Trooping by air was now an accepted part of military life, and the majority of service personnel appreciated the Hastings speed of passage to distant places - although a few masochists did complain of over-feeding, whilst other poor souls hid their misery in paper bags!

On 1 September 1958 511 Squadron was renumbered 36 Squadron, then on 5 May 1959 114 Squadron was reformed at Colerne with

HANDLEY PAGE HASTINGS AND HERMES

C.Mk.1A TG527 of Transport Command. The underwing fuel tanks were non-jettisonable! [R. Deacon]

Hastings. In September of the same year the Hastings T5 variant was introduced, this was the conversion of Mk.1 aircraft for use with the Bomber Command Bombing School at Lindholme, to train the V Bomber crews on electronics and Vulcan H2S radar. The T5 aircraft formed '1066 Squadron', whose C.O was Squadron Leader Jackson, who complimented the Hastings thus: *"Another thing to do with the Hastings, which is my favourite,I think I enjoyed flying it more than any other".*

This unit and its Hastings T5s would in the end be the ones continued with, when the other squadrons had been converted to Britannia and Hercules aircraft. Some of the aircraft in use by the BCBS at this period would conjure up memories for the early members of Hastings squadrons, aircraft such as TG503, TG505, TG511 and TG517. Training flights usually lasted 4-5 hours and provided training for a number of trainee radar operators on each flight. Flight Lieutenant G.Heath trained on them during the early 1960s as a radar operator for the Vulcan force, said of these remaining members of the Hastings fleet:

"I remember them as reliable aircraft that did a successful job as a training aircraft. Being a navigator I have no experience of the handling characteristics, except that they gave the rear crew a reasonable ride, even though they had been severely modified to take the radome required by the NBS." The Hastings squadrons were still well occupied with trooping, supply and paratroop dropping execises, as well as being on call for any emergency or back-up to the Bomber Command units.

The aircraft in the U.K were still assigned to the station and not the unit, although the crews were. So 1960 commenced with Hastings forming the equipment of 24, 36, 48, 70, 114 and 202 Squadrons, with 70 Squadron still based in Cyprus and 48 Squadron at Changi, Singapore. Before C Mk.1 aircraft were re-issued on re-equipment or when due for refurbishment, they were not only brought up to the latest modification standard, but their fuel tank capacity increased by the addition of a non-jettisonable fuel tank under each wing and Hercules 106 engines were installed. These aircraft were then designated Hastings C.Mk.lA, they were then more capable of long-range operation. Flights from Lyneham to Changi taking five days, unfortunately, neither these nor the VIP version were pressurised.

About this date a number of C Mk.1 aircraft were struck off charge and disposed of to a metal salvage company. A further Hastings was disposed of on 1 March 1960, when TG579 of 36 Squadron was flown too low on the final approach to Gan runway and finished in the sea, fortunately without casualties - except to the pilot's pride.

1961 saw another confrontation commence, when Iraq began offensive manoeuvres and belligerent noises against its smaller neighbour Kuwait. This resulted in the ruler of oil-rich Kuwait calling on Britain to honour its defence treaty. Following this a massive airlift commenced, with aircraft flying in supplies and personnel, which included elements of the Strategic Reserve. Trooping aircraft were routed in from all parts of the active Commonwealth, this included most of the Hastings squadrons. Hastings took part in shuttling troops from Nairobi in Kenya to Kuwait, the heat at the Kuwait end being so hot and dry that it was estimated that RAF personnel alone consumed 50,000 bottles of soft drinks and 7,000 gallons of water during the emergency.

The Hastings aircraft were prepared for supply dropping as well as for the trooping of personnel, but lack of air conditioning and pressurisation, the temperature inside the aircraft was intense and the metal skin whilst standing at Kuwait hot enough to instantaneously fry an egg. One long-serving airman, ex-North Africa, claimed that service in Kuwait should qualify for double overseas service!

The biggest problem with the Hastings in the

SERVICE USE OF THE HASTINGS

Hastings TG610 of 242 OCU crashed on landing at Thorney Island on 17 December 1963 - this was the end result! [P. Porter]

freighting role was the one applicable to all tailwheel undercarriaged aircraft, that of loading the freight - whether it was boxed, bagged, jeeps or artillery - the movement of it uphill inside the fuselage. It was a major point of dislike by Army personnel when loading their military equipment, and Army representatives at conferences were to stress this point a number of times, ignoring the amount of mobility that the Hastings had conferred on them - didn't pongos always have to complain?

Although it may appear by this text that incidents and accidents dominated the Hastings history, the facts are quite different, but because it was a transport aircraft it was without the glamour publicity that was attached to strategic bombers and sweeps by combat aircraft, and its routine flying across the world's surface to areas of conflict gained no kudos from the Press. Thus, when an incident or accident did occur this did tend to get highlighted rather than the miles and miles of safe operation. In the light of its operating over the years it would be difficult to determine which was the most impressive part of the Hastings daily routine, the many places it operated to or over in the course of its world-wide duty, or the many roles that it so successfully accomplished during its years of service.

One incident on a routine flight from Christmas Island back to its base at Colerne raised no world headlines, but merited sixty lines of newsprint in the local Paper. A 36 Squadron Hastings crewed by Ft/Lieutenant Jackson, Flying Officer Franks, Master Engineer McEwan, Ft/Lieutenant Hughes and Sergeants Allan and R.Smith were called in on an emergency to search for a Japanese cadet training ship, which was reported drifting and sinking. Failing to find the vessel in the reported position, the Hastings crew commenced a search pattern and after a further hour of searching eventually located the vessel, the ship's crew already prepared to abandon ship. The Hastings radioed the position of the ship and remained transmitting until an American rescue aircraft homed on them and took over the rescue.

When trouble in the ex-Belgian Congo became the problem of the United Nations the Hastings of 24 and 36 Squadrons became the heavy lift component of the British contribution. Flying from the U.K to Accra in July 1960 they were employed from there to fly Ghanaian and Nigerian troops and supplies of the United Nations to the Congo. On one Hastings' flight list to the Congo there were Army officers from twelve different nationalities. The crews were based at either Lagos or Accra in Ghana and flew into either Leopoldville or Katanga, the latter's residents were on the opposite fighting side to Leopoldville. This placed the aircrew in a most invidious position as they were flying over the front line, and incidents on the ground did sometimes occur due to ignorance on the part of the local native troops.

The usual procedure for the positioning of the Hastings on this operation was to fly to West Africa, carry out their transport duties there and fly off 300 hours, then to return to the U.K for the major inspection. This of course entailed a fine bit of timing towards the end of the period to allow for the time of the return journey to base. The return journey was usually by direct non-stop flight from Accra to Idris (Tripoli), a flight time of around 10 hours. If the aircraft had flown outbound with passengers and freight, then there was always the possibility of no ballast being carried for the return flight. Whilst this posed no problem during the flight, a problem did arise when the power was cut on landing, as a nose-down attitude was adopted by the aircraft. To counter this the procedure was for the captain to give the flight engineer the order "Very slow cut". Failing to carry out this order slow enough resulted in the Hastings taking the bit in its teeth and putting its nose down smartly, hence the remark *"I thought I had mastered the b******, then it turned and bit me."*

So the normal procedure when flying on crew training or without freight or passengers, was to carry 1200 lbs in ballast on the C Mk.1 and 1000 lbs on the C Mk.2. After take-off from Idris (or Castel Benito) the flight was direct to Lyneham, Customs, a rest, then another assignment - the world was the Hastings crew's oyster.

HANDLEY PAGE HASTINGS AND HERMES

Underside view showing plan form of C. Mk.2 WJ340 of 24 Squadron Air Support Command. This aircraft was eventually struck off charge on 7 February 1968. [Sqn. Ldr Jackson]

By June 1961 there were only six units flying the Hastings in Transport Command, 24, 36, 48, 70, 114 and 202 Squadrons, the aircraft by then mainly operating as a main line trunk service. Yet, even after the entry into service of the Britannia, Comet and VC10, the Hastings still managed to perform the same function - though more leisurely than its younger brethren, and came into its own on short-range operations in Europe, the Middle East and Far East.

One experienced pilot of long standing was P.F.'Red' Eames* (* Veteran of the Battle of France 1939-40 flying Battles and Blenheims. Post-war flew Sunderlands and Shackletons. Retired as Squadron Leader.) who flew the Hastings while attached to A&AEE Transport Flight, and flew on it to both Canada and Australia. In a letter to me:

"It took me some time to get used to the elevator control, which I found stiff and high geared. At the same time, I recall that the aeroplane was a steady enough platform on instruments, particularly in the cruise, and I found it easy enough to squeeze an extra couple of knots when all was quiet.

"One developed a respect for the aeroplane as a good, honest performer, and the Transport crews didn't have any moans that I can remember. How much nicer it would have been on a tricycle (undercarriage) - I think that you will find as I did, that comparison with the contemporary DC4 showed the old Hastings was superior in speed,

height and load carrying performance, but didn't get any kudos. It was certainly the first aeroplane that I had flown which could cruise at 200 knots true over a 2,000 n.m sector with any sort of load"

The Hastings also achieved spectacular results when transporting squadron personnel around the U.K and Germany, and in particular on what were known as 'May flights', this was the activation of Bomber Command 'V' Force when dispersing to its wartime bases. The 'V' Force crews, who took hours to brief, could not understand how it was that Hastings crews could be in bed, get called out, arrive on their station and transport them elsewhere in a very short time-scale. The secret lay in the scheduling, preplanning and familiarity with the whole area.

Secrecy also lay behind the Special Duty Flight, this was in existence during the hey-day of the Hastings. The personnel of the Flight were not answerable to the Station Commander in general, but took their orders from London or Command Headquarters. When the order came they and their aircraft disappeared, sometimes for long periods of time - but what they did and where they went still remains a secret, indicating the security and integrity of the crews. As one Hastings aircrew member pointed out to me, if the Hastings was good enough for special flights, then its reliability could not have been in doubt. Maybe the snide remarks about "Three engines" had got worn out, the old warrior had settled down, but the Hastings

SERVICE USE OF THE HASTINGS

Hasting starboard outer engine having it's plugs changed in cold surroundings. [P. Porter]

aircrew speak well of her and her strong construction - even if not all landings were 'greasers'.

Route flying with the Hastings, as with the York and previous aircraft, was comparable with the relaxed days of the flying boats, in that it normally operated on the route during the day and nightstopped en-route. This was due to the fact that it had no weather radar and that it was more economical to nightstop with all the various staging posts available to the RAF throughout the world. At the same time, the long-range capability of the Hastings did allow for over-flying, or avoidance of, countries in disagreement with Britain's policies.

The end of 1964 saw only four units left operating the Hastings, these were 24, 36, 48 and 70 Squadrons, 202 Squadron Coastal Command having relinquished their Hastings in July 1964, but had operated their aircraft over fourteen years without a fatal accident. 51 and 99 Squadron were each operating one Hastings as part of their complement. One member of 48 Squadron was D.P.Hibert, who remembers the Hastings with affection and feels that most people who worked on the Hastings felt that way. Said of the aircraft: "It is 20 years since I was involved with the Hastings, or Hasti-bird as most of us referred to it. I cannot recall any real problems ... to my mind the aircraft was easy to service and the servicability rate was very good, the only problem I recall was due to water in the fuel tank even after the water drain check. Due to the attitude of the aircraft on the ground it was possible for water to remain in the rear of the tanks behind the fuel outlet.

"My favourite story relates to the friendly rivalry between 48 Squadron flying Hastings and 215 Squadron flying Argosy aircraft, as we had the older aircraft much fun was made of them. Our response was in verse, unfortunately the wording long forgotten, however, the gist was that the Hastings could fly to Gan with a full load and the Argosy could just about make Gan with a load of ping-pong balls!"

Over a number of later years the Hastings had suffered fatigue problems in the area of the elevator hinges and their attachment, all have

A Signaller at work in a Hastings, using a STR18 transmitter/receiver. [P. Porter]

HANDLEY PAGE HASTINGS AND HERMES

been well documented and relate specifically to the attachment bolts, and all were covered by modification action or special inspection notices. Apart from the one case where the modification had not been incorporated, there appeared to be some confusion on bolt tightness. Official investigation gave the impression that some of the fatigue was due to the elevators vibrating under airflow from the engines on the ground with the elevator locks 'in', but what about the Hastings that were used for paratroop or supply dropping, when considerable time was spent at low speed with the flaps down, causing turbulent airflow over the tail and elevator surfaces? The surprising point of the checking of the bolts, was that when a full check was carried out of all bolts there was no set pattern of failed bolts, their position numbers or length of service.

Then on 6 July 1965 there occurred the worst Hastings accident, when TG577 of 36 Squadron took-off and approximately one minute later the pilot reported to the ATC that he was having "trim trouble" and "controls sloppy" and requested a priority landing. The aircraft was seen flying a wide circuit at 1,500 feet. It was then seen to stall and dived into the ground, killing the crew of six and thirty-five Army and RAF parachutists on board. TWO fractured elevator bolts being found near the wreckage.

It was determined that the aircraft had flown 7,416 hours since new and that the tailplane had been overhauled in September 1960, when new bolts had been fitted. Since that date the aircraft had flown 2,411 hours and all modifications and inspections had been carried out. Two days later all the eighty Hastings still in service were grounded. The Court of Inquiry concluded that the root cause was the failure through fatigue of the two upper bolts attaching the outboard elevator outrigger to the tailplane drag member. An RAE report was produced covering the metallurgical examination of the failed bolts. The Court of Inquiry suggested that when remedial action was taken that emphasis was given to correct fit and alignment of nut and bolt in its assembly, and that a plain steel washer was fitted under the bolt-head and under the nut. Following this there were no further incidents concerning failure in this area.

Following this the Ministry of Defence then announced that before any Hastings was returned to service all modifications and inspections were to be carried out on the whole tail unit and the elevators re-conditioned This work was undertaken by Handley Page in the U.K. and Maintenance Units overseas.

In 1965 four Hastings were flown from Lyneham to Nassau on detachment, their duty being to fly daily to Belize and back, either carrying replacement troops or supplies. The return flight carrying out the relieved (in more senses than one!) troops. The stationing of British troops and aircraft in Belize had come about due to Belize's larger neighbour, Guatamala, making territorial demands, so Britain was honouring a defence agreement with Belize (originally British Honduras).

The flight plan from Nassau to Belize was such that Cuban air space was avoided, and so the flight was made along the Florida coast and then across the Gulf of Mexico to Belize, making position reports to Miami, New Orleans and anyone else interested in conversing, using a call-sign of 24 hours duration - this latter affects the story that follows.

On 6 April WJ330 took-off from Nassau flown by Flying Officer 'Spike' Marie, with Maurice Patterson as signaller in the crew. The aircraft proceeded on its Belize shuttle and made position reports en-route. Before the destination was reached it became necessary to feather the starboard outer propeller due to engine temperature fluctuations. Upon arrival at Belize it was decided to night-stop to allow the groundcrew time to fix the snag. However, when the following day dawned WJ330 was in no better shape and still only powered by 3 engines. As the airfield had only limited facilities and accommodation - described in ruder terms by the troops - it was decide that the flight should be made lightly laden and only the troop's equipment loaded.

Once airborne and on the way the engine again began misbehaving, so the propeller was feathered and the flight continued as normal - for a while anyway. Maurice Patterson:

"I started into my positioning reporting routine, both to the civil authorities and the USAF (using the allotted call-sign), but we had forgotten that it was only of 24 hour duration, and the next day would be allotted to the other aircraft".

Result was two Hastings crossing the Gulf of Mexico in opposite directions at different heights and slightly differing tracks, but both with the same call-sign - panic in SAC control ensued.

"Instead of immediately noting that the two courses were nigh on reciprocals, they latched onto the call-signs only and wondered how in hell two big gooney birds could move so far, so fast. At the next hourly report I had to give practically a life history of the crew and the trip, and was advised, nay, ordered to remain on that frequency. What happened next we found highly amusing. Suddenly(!), like the voice of God there boomed in our earphones a sepulchral "Air Force 330", so I jumped up to look out of the astro-dome, and there about 300 yards astern and slightly above were a pair of USAF fighters. After a concise cross-examination, we found that our antics across the Bay had caused a minor alert in SAC (Miami) area. Once these boys were satisfied the leader decided to formate on our starboard side close in. To do this he found he had to drop his wheels and flaps, upon which Spike slowly eased off the power to about 120 knots IAS and the leader fell out of the sky in a stall and everyone on the frequency learnt a new four letter American expletive!"

The Hastings in the remaining squadrons continued with their routine work in support of the Services, providing medical evacuation, transportation of supplies and equipment, and still tramping to various parts of the globe. Its

SERVICE USE OF THE HASTINGS

long-range capability making it possible to over-fly or by-pass newly independent nations who disagreed with British action or policy in another part of the world, areas that Britain was still responsible for or held agreements with. By 1967 however, the Lockheed C130 Hercules began to arrive in squadron service and the Hastings was progressively withdrawn from transport operations. The last unit to relinquish its Hastings was 24 Squadron, which it did on the 5 January 1968, the Hastings had completed twenty years of service in Transport Command.

RNZAF Hastings.
The Prime Minister of New Zealand, S.G. Holland, received the RNZAF's first Hastings C Mk.3 from Sir F.Handley Page at Radlett on 30 January 1952. The aircraft was NZ5801, the first of four. This model was powered by Hercules 737 engines and was slightly faster than the RAF's C1 and C2 aircraft - the RNZAF crews were not slow in demonstrating this, with suitable gestures! Their main duties were to be as troop transports and ambulances, as well as long-range freighters. In this respect their first major airlift was the transportation of No.14 (fighter) Squadron RNZAF to Cyprus. The aircraft formed the Hastings Flight of 40 Squadron, but in 1953 it was re-numbered 41 Squadron.

The fourth aircraft NZ5304, was entered in the London to Christchurch Air Race, the crew being led by Wing Commander R.F.Watson. Everything went well until approaching Negombo (Ceylon) from Masirah, when the No.2 engine lost power, so its propeller was feathered after engine shut-down. At this time Negombo was covered by a tropical storm and NZ5804 was plowing through a monsoon and about one hour from touchdown. On landing on the heavily flooded runway at Negombo water was thrown up in sheets from the main wheels and damaged the flaps. Nevertheless, it was decided to carry on without flaps, but upon investigation of the No.2 engine filters it was found the engine had failed, which if course necessitated withdrawal from the race.

N5804 appeared to be fated, for it was lost during its first year of operations, when during a take-off at Darwin it suffered an engine failure, the pilot re-landed on the runway at 130 knots, but failed to stop. The aircraft tore off the runway, across a ditch, up a hill, over a road, through a twelve inch diameter pipe and on to a railway track. The strength of the Hastings was proved again, for although it was damaged beyond repair, the aircraft neither burnt nor broke-up and no one was injured. It was jokingly remarked in Darwin that this one Hastings had done more damage to Darwin than the Japs had done in WW2!

With the introduction of the Hastings into service, the aircrew training programme kept the first aircraft in the air 51 hours over five days. Throughout their service with RNZAF the Hastings were maintained at a high rate of utilisation, and it was claimed that by the end of its service that not one scheduled flight was ever cancelled for want of a serviceable Hastings. The RNZAF with its Hastings must have operated the longest air route in the world at that time, for it flew from its base at Whenuapai (North Island) to Lyneham, a distance of 13,750 miles.

When 14 Squadron RNZAF fighters were stationed in the Middle East in Cyprus, the squadron's aircrew and ground personnel were airlifted by Hastings from New Zealand. Then, during a five week period they carried mail, supplies and personnel in return flights from New Zealand to Cyprus and on to the U.K, covering approximately 140,000 miles. During this operation - a quick 'turn-round' was usually made at base, a matter of two days at Whenuapai, before the aircraft returned back along the route.

Another regular service for the Hastings was the 1,300 mile flight from Whenuapai to Nandi in the Fiji Islands, supplying the flying-boat squadron based at Lauthala Bay. Stores and mail were also dropped to weather stations and small farming communities in the Chatham and Kermadec Island group. These lay off the main steamer routes and so contact was rare, the Hastings provided a means of supply. The 'drops'

Hastings C.Mk.3 NZ5801 of the Royal New Zealand Air Force awits collection at Radlett [J. Knivett]

HANDLEY PAGE HASTINGS AND HERMES

C.Mk.3 NZ5802 of the RNZAF at Khormaksar, Aden during October 1961 [B. Gardner]

were usually made from 50 to 60 feet. In some cases the DZ on the island was small, as well as the island being small, so not only did navigation have to be accurate, but the dropping of the supplies as well.

During August 1953, one Hastings on a supply dropping mission, had a large packet foul the tailplane, causing damage that was sufficient to create partial loss of control that necessitated a forced landing on a 5,000 foot grass strip. Seven days later with the damage repaired, the Hastings was flown out successfully by the same pilot.

The Korean War opened up a new route for the Hastings, which were used to fly New Zealand troops to Japan, ferrying personnel and supplies both to there and other bases, returning with casualties or time-expired troops. Ocean searches for missing ships were carried out, relief supplies to isolated islands as well as transportation of sick residents to hospital were all part of the Hastings' tasks. One Hastings made an eight day tour of Australia for liaison with the RAAF and to give demonstrations. The Hastings also enabled senior officers of the RNZAF to meet their opposite numbers in the RAAF, and for the aircraft to be demonstrated to the School of Land/Air Warfare at Williamtown in NSW, Australia.

In 1956 NZ5803 captained by Squadron Leader K.B.Smith was flown on a VIP mission to the USA, this was to take the RNZAF Chief of Air Staff, Air-Vice-Marshal C.Kay, to Bolling Field at Washington. On arrival at Bolling Field after an uneventful flight it was found that a 20 knot cross-wind was blowing, which kept the pilots fighting to keep the aircraft on the runway, which they did, and the mission was successful, although again a tailwheel Gooney Bird did raise a few eyebrows.

The Hastings in the RNZAF began to be replaced in 1964 with Douglas DC6 aircraft, these pressurised aircraft being purchased from TEAL, but the Hastings continued to operate regularly alongside the new aircraft until eventually they were retired on 2 February 1966, when they were replaced by Lockheed C130 aircraft. By then the RNZAF Hastings had flown 7,106,000 miles in 29,003 hours, and during this period had not incurred one single casualty. They had flown one VIP flight to the USA, 179 flights to Singapore, 133 flights to Fiji and the Pacific Islands, 50 flights to the U.K, 29 flights to Australia and seven flights to Japan. During its service with the RNZAF the Hastings suffered very few troubles and few delays were incurred through unserviceability. Its structural robustness was legendary - as was its desire to fly on three engines in the early days. Yet by the time of its retirement from service the Hastings had proven reliability and was viewed with affection, even though classed as an anachronistic piece of hardware by some. Yet for take-off there was none of the modern need to consult take-off charts, all that one needed to do was grab a handfull of throttle levers, push them forward to the stops, and four Hercules engines would life the Hastings into the air with ease. The last four piston-engined British built transport aircraft in the RNZAF.

Far East.
Blackburn Beverley aircraft had by August 1957 begun to arrive to equip U.K based squadrons, and so Hastings aircraft therefore became available to re-equip overseas squadrons, amongst whom was 48 Squadron. The unit was based at Changi airfield Singapore and had been operating Valettas. The primary role of the squadron was the provision of air transport to the Far East land forces (continuing a role that had been in existence from when the first bomber-transports had arrived pre-war), and to provide scheduled services throughout the area carrying passengers and freight. Over the years it would add further operational roles to its repertoire, amongst these would be the involvement in the operations at Kuwait, as well as a four year detachment to Christmas Island in the Pacific for the British nuclear tests.

On 30 June 1961 the ruler of the oil-rich sheikdom of Kuwait, which had a defence agreement with Great Britain was threatened by a

SERVICE USE OF THE HASTINGS

powerful invasion from Iraq. So the Sheik called on Great Britain to honour its agreement. A large airlift was inaugurated and amongst the Hastings taking part were six aircraft and twelve crews from 48 Squadron. The unit moved into Khormaksar in Aden as a temporary base and began to airlift troops and equipment from Nairobi in Kenya. Servicing at Kuwait was carried out in temperatures sometimes reaching 130°F, so that heat even tested the servicing crews used to Far East temperatures - thirsts as well as liquid consumption was high! Inside the Hastings' fuselage the temperatures were unbearable and where possible flying commenced early in the morning. Tension in the area decreased during July, so that by the 21st of the month the aircraft were beginning to depart to their home stations.

48 Squadron also maintained a detachment at Kuching for supply dropping duties and dropped 14 million pounds of supplies. The squadron lost WD497 on 29 May 1961 during a practice supply drop at Seletar, an engine 'cut' and on the turn-in the aircraft dived into the ground; the accident report considered that the pilot lost control in asymmetric flight. A detachment was also maintained at Butterworth to supply jungle forts and airstrips, giving succour and support to Army patrols and army/air support where necessary. Some of the Hastings operating in the Far East were TG614, TG620, WD481, WD488, WD499, WJ332 and WJ333; their colour scheme being silver below a cheat line and white above it.

In 1962 a confrontation with Indonesia commenced, this occurred because of fierce opposition from President Sukarno of Indonesia to Britain's concept of an independent Greater Malaysia, which was to incorporate North Borneo, Singapore and Malaya. The opposition stemmed as much from Sukarno's desire to acquire oil-rich North Borneo as it did fron jealousy of a possible larger neighbour. The action began when insurgents in the British protectorate of North Borneo staged an armed rebellion against the Brunie civil authorities. These rebels were the TNKU, who were in sympathy with the Indonesians' aspirations to incorporate North Borneo into a Greater Indonesia. Labuan Island was 120 miles off the coast of Borneo and was an RAF staging post, so provided an intermediate base for any action or reinforcement of the area.

In December, 'Operation Borneo Territories' was commenced to give aid to the Brunei authorities. Amongst the first aircraft air-lifting two companies of the Gurkha Rifles to Brunei were twelve Hastings of 48 Squadron. By the end of 8 December the rebels had used their initiative to virtually capture the whole of Brunei, so further loads of troops and supplies were flown in on the 9th, and this continued over the following week. In the end, five transport squadrons were involved in this operation, and of the 2,000 sorties flown 48 Squadron Hastings flew 1,000 of them.

Although the initial and local revolt was crushed during early 1963, raids were still being carried out across the border from Kalimantar, that in the end required the setting up of posts near the border and local patrolling. This gradually drifted into the Indonesian Confrontation due to Indonesia support of the rebels, and Indonesia breaking off diplomatic relations with Britain and the new state on 16 September 1963, when Greater Malaysia was established. Three days later three Argosies of 215 Squadron and one Hastings of 48 Squadron,

KJ333 - a Hastings VIP aircraft of Far East Command at Sandakan, North Borneo in 1964 [M. Pattison]

HANDLEY PAGE HASTINGS AND HERMES

C. Mk.1 of 48 Squadron carrying out a re-supply drop in the Far East. [M. Pattison]

under a guarantee of non-interference, flew into Djakarta and evacuated the British community to Singapore.

Labuan and Kuching were the two forward airfields into which supplies and personnel were flown into Sarawak, and from where supplies were flown to the forward areas. These areas also included forts built on stilts in various valleys, through which the Indonesians would be forced to infiltrate if they wished to mount an incursion. So to these forts it was necessary for the Hastings to make their approach through the valleys and thus in range of small-arms fire. There were reports of one navigator having his cup of tea shot away just as he was going to drink it, and a flight engineer found a bullet embedded in one of his throttle levers.

This confrontation continued in a defensive mode for a while until in the end Britain began mounting raids and ambushes within Indonesian territory. Eventually, peace of a sort returned to the area and the Indonesian Government were forced to accept Greater Malaysia as an established fact. Meantime the Hastings continued with their supply dropping along with other aircraft, this supply sometimes resulted in the 'accidental' crossing over the border during the dropping pattern. One story has it that after a period of supply dropping at one fort the Army picked up an Indonesian NCO deserter. Apparently, as the story goes, after three days of supply-dropping aircraft over-flying an Indonesian anti-aircraft post, its Company Commander gave out the order that *"The next big aircraft that does that, you shoot - or else!"* So on the next appearance of a large four engined aircraft the gun-crew opened fire and shot it down unfortunately it was an Indonesian Air Force Lockheed C130, hence the desertion.

During 1962, 48 Squadron also had to provide a two Hastings detachment to Christmas Island in the Pacific for the nuclear bomb tests, the guard for the bomb components never relaxing, even at night-stops. The Hastings were later joined by an American detachment during the bomb tests - the USAF personnel viewing the Hastings as a relic from another age, failing to realise that it was as fast as their DC6s and Constellations. On 17 January 1963 Air Chief Marshal Sir Walter Merton presented 48 Squadron with its Standard on behalf of the Sovereign; the squadron at this time still providing support to the land forces, as well as food and medical supplies to the aborigine settlements in Borneo.

During the temporary grounding of the Hastings in 1965 after the serious accident at Abingdon, inspections were carried out on the Hastings in the Far East at Changi. Three aircraft were found with faults, these were TG614 with four out of six rivet heads missing from the lower bracket on hinge rib of port outboard outrigger; WJ333 with the starboard inboard bracket fractured at top; and WJ336 with its port inboard elevator bracket cracked.

The 48 Squadron detachment at Christmas Island had the task of carrying out flights to Honolulu to bring in fresh food and fruit supplies as well as mail, not to mention giving personnel a break away from the monotony of Christmas Island life - so as can be imagined, any technical breakdown or any hold-up at Honolulu was viewed with favour! The Americans there nicknamed the Hastings the "Four-engined Gooney Bird" to differentiate it from their own "Gooney Bird", the Douglas C-47, so namewise it was in good company.

In February 1967 came the temporary disbandment of 48 Squadron and their break with the Hastings. On the disbandment of the squadron at the farewell parade, the following was

SERVICE USE OF THE HASTINGS

A number of transport Command Hastings at Christmas Island as part of 'Operation Grapple' task force. [M Kennedy]

said: "... *farewell to the faithfull Hastings, which has served so well for ten years*".

With the disbandment of the squadron some of the Hastings and personnel were transferred to the Far East Communication Squadron at Changi. This included one VIP aircraft, which was considered to be more VIP than the standard C4 VIP aircraft, described by one crew member "... as having all 'mod cons' bar the kitchen sink..." but he did not mention a well-stocked bar, so we can only surmise! This particular aircraft was also fitted with STR18 radio equipment as opposed to the older Hastings, which still had the TR1154/155. With the STR18, fantastic freak results were sometimes obtained in the Far East, the radio officer often being in touch with Changi when at 10,000 feet over Saigon, yet minutes later losing contact until within normal range. The Communication Squadron with their Hastings also carried out calibration flights to places like Gan, but this was with the standard aircraft and without the delights 'suffered' on the VIP aircraft

The squadron continued operating the Hastings for about a further year, when upon relinquishment the aircraft were flown 8 to 390 M.U at Seletar and broken up. The crews were in no doubt about the Hastings, considered it "...*a docile and forgiving aircraft in the air*", and although there were still a few critics who would call it "The best three-engined aircraft in the

Hastings Met Mk.1 TG517 of Coastal Command in the smart new livery. .

HANDLEY PAGE HASTINGS AND HERMES

world", by then that period was long passed, and crews pointed out that even on three-engines it could out-perform comparable aircraft of its day, so using the same expression as a back-handed compliment.

Experimental conversions.
Well-known at RAE Farnborough and also based at Farnborough was C Mk.2 WD480. It arrived at Radlett from Aston Down on 3 July 1951 for conversion to carry sonobouys for special trials. The alterations rendered the aircraft unsuitable for use in the various roles that Hastings were employed on. An understructure was constructed similar to a pannier, built-up from aluminium alloy on heavy frames forming bay bulkheads with light frames inter-spaced between them. This structure not only provided stowage for twelve sonobouys that was covered by bombbay doors but also carried a scanner radome. The frames were riveted and bolted to the lower surface of the fuselage and centre-plane, the loads being transferred to the basic structure of the airframe by additional fittings. The fuselage interior was also different in that the cabin was divided into two compartments, with the forward one equipped as a laboratory and the rear one provided space for technicians, a launching tube and a rack of marker bouys. The sonobouy bay doors were operated by hydraulic power controlled by selectors on the pilot's control pedestal (drawing 11).

WD480 arrived at Farnborough as part of the Ministry of Supply Fleet on 1 April 1953 and was allocated for sonobouy trials. Over the years it was used for long-range sonobouy trials, aerial development and special trials, arriving back at Radlett on 30 August 1966 for reconditioning and the installation of Hercules 216 engines. It was allotted back to RAE Farnborough and was eventually struck off charge on 25 September 1974.

Since 1947 the U.K had operated a Meteorological Research Flight from RAE Farnborough, the Flight being part of the Research Division of the Meteorological Office. Hastings TG619 joined the Flight in 1955 and had been extensively modified and instrumented. It was further modified to eliminate problems and for experiments. The Flight operated two main operational flights, a low-level one up to 22,000 feet and the high level one above this altitude, the Hastings being assigned to the lower level flights. Measurements taken covered icing, air temperatures, water content of clouds, humidity and the investigation into the visibility in rain, air turbulence and vertical currents. TG519 also took part in trials of a number of radar, anti-radar and homing devices over the years 1955-1963.

A large number of Hastings were used over

Above: Hastings C.Mk.2 WD480 of RAE Farnborough with built-on pannier and radome. [MAP]
Below: Hastings C.Mk.2 TG502 used by A&AEE Boscombe Down.

SERVICE USE OF THE HASTINGS

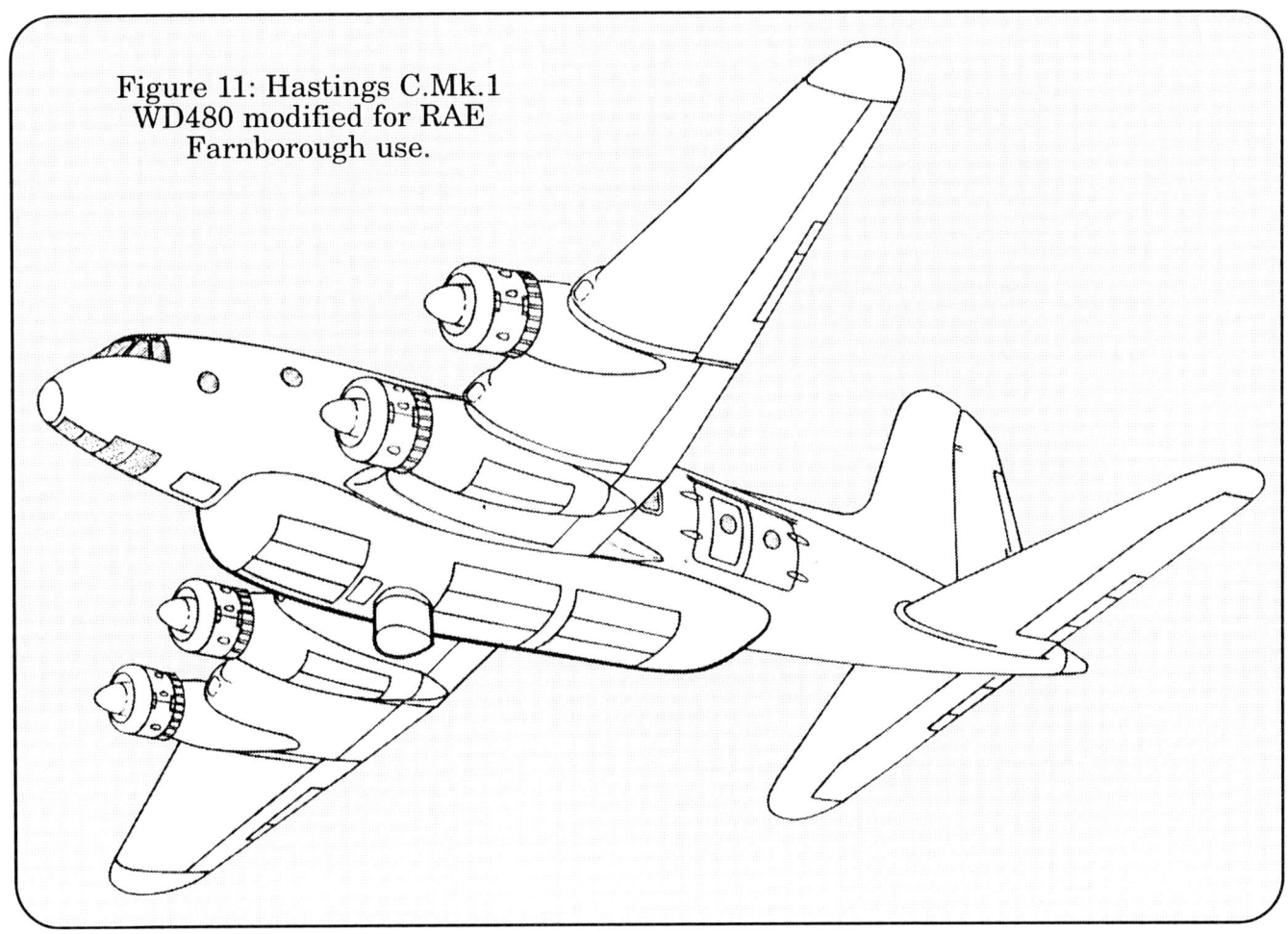

Figure 11: Hastings C.Mk.1 WD480 modified for RAE Farnborough use.

the years at various research establishments, these included TG500, TG501, TG502, TG503, TG506, TG514, TG533, TG580, TG619 and TE583. This latter aircraft was originally allotted to A&AEE on 17 January 1947 for handling trials, then allotted to Handley Page at Radlett for its use as an engine test bed. In this role it was installed with Armstrong Siddeley Sapphire gas turbine engines in the outer engine positions with the necessary instrumentation and auto-observers in the main cabin. The trials were to prove the operation of that type of engine for use on the Victor bomber. On 30 May 1951 TE583 departed Radlett for Bitteswell to enable the NGTE to monitor and evaluate the Sapphire engines under flight conditions. By May 1952 it was again at Handley Page for the installation of a Victor type escape door and markings were painted on the rear fuselage for use in the evaluation of the escape door. In 1954 it was posted to RRE for use in the development of many types of radar, including the development of the sideways looking radar for the TSR2, and was eventually released from service in 1965 and during the year it was scrapped at Pershore.

TG500 arrived at Boscombe Down on 27 April

Hastings WD499, used by the Radar Research Flying Unit of the Royal Aircraft Establishment. Note the two built-in long containers under the fuselage. [Author]

HANDLEY PAGE HASTINGS AND HERMES

1948 after the incorporation of a number of TI modifications, and was despatched back to Handley Page within two months for further modifications. On 17 January 1949 it was being prepared for heavy dropping trials and served at A&AEE and AFEE, following which it received a nose radome and was allotted to A&AEE Boscombe Down. Amongst a number of tasks carried out were micro-wave radar tests for the MCA and was eventually struck off charge on 12 April 1973.

Hastings C2 WD499 joined the Radar Research Flying Unit of the RRE and was tasked to carry out radar research programmes as well as a number of other projects. Amongst the latter was a joint air-sea exercise during September 1972 run by both British and American scientists. For this the aircraft carried Loran, Omega and a number of special sensing devices. Some of the latter were dropped into the sea at pre-set points and transmitted back the sea temperatures. Under the fuselage were two narrow structures, one each side. The life of WD499 ran out on 30 September 1974 when on retirement it was transported by road to Honington for use in fire training.

Training role.
As explained earlier the conversion training on to the Hastings aircraft was carried out by 242 OCU, initially at Dishforth, where the pattern of training was set out. The Hastings aircraft that were operated by the OCU were equipped with a quite comprehensive radio and radar fit, having ARC 52 UHF transmitter/receiver, TR1998 VHF transmitter/receiver, STR18 MF and HF transmitter/receiver, radio compass, AYF radio altimeter, ILS, 1961 amplifier for intercommunication, USA Command receiver, Rebecca Mk IV, Gee II and IFF Mk.10. For long-range route flying in certain parts of the world the Gee II indicator was removed and a Loran indicator fitted instead.

Early in 1962 No.2 Air Navigation School (ANS) moved out of RAF Thorney Island and was in February replaced by 242 OCU with their Hastings and Beverleys. Apart from the straight conversion role, 242 OCU aircraft were quite often fitted out for various roles, this included passenger carrying, cargo carrying, paratrooping and supply dropping. The latter entailed the fitting of roller runs, two side by side with a three to four foot gap between them, running the full length of the fuselage and at the last few feet curving at an angle to the entrance door. Unwary aircrew or groundcrew often found that accidently treading on the rollers could mean a hasty exit, not an entrance!

TG508 on 7 March 1962 made a bad landing, skidded across the grass and was brought to rest by a mound of earth. This was fortunate, as the aircraft was heading for the hangar, so there was the makings of a disaster. The crew evacuated the aircraft promptly, people spoke of the navigator complete with nav-bag, popping out the astro-dome hatch like a cork out of a bottle. Then the aircraft caught fire, with fuel spreading under the PSP, requiring in the end the help of the Chichester fire brigade to provide extra foam to help out the Station Fire Section.

With so much flight training locally, groundcrew had the opportunity to have flights in the Hastings, especially if it was to help in the clearance of a flight snag. The late Warrant Officer Paddy Porter relates one of these when he was a Junior Technician:

Nose view of Hastings T5 starting it's engines.
[P.Porter]

SERVICE USE OF THE HASTINGS

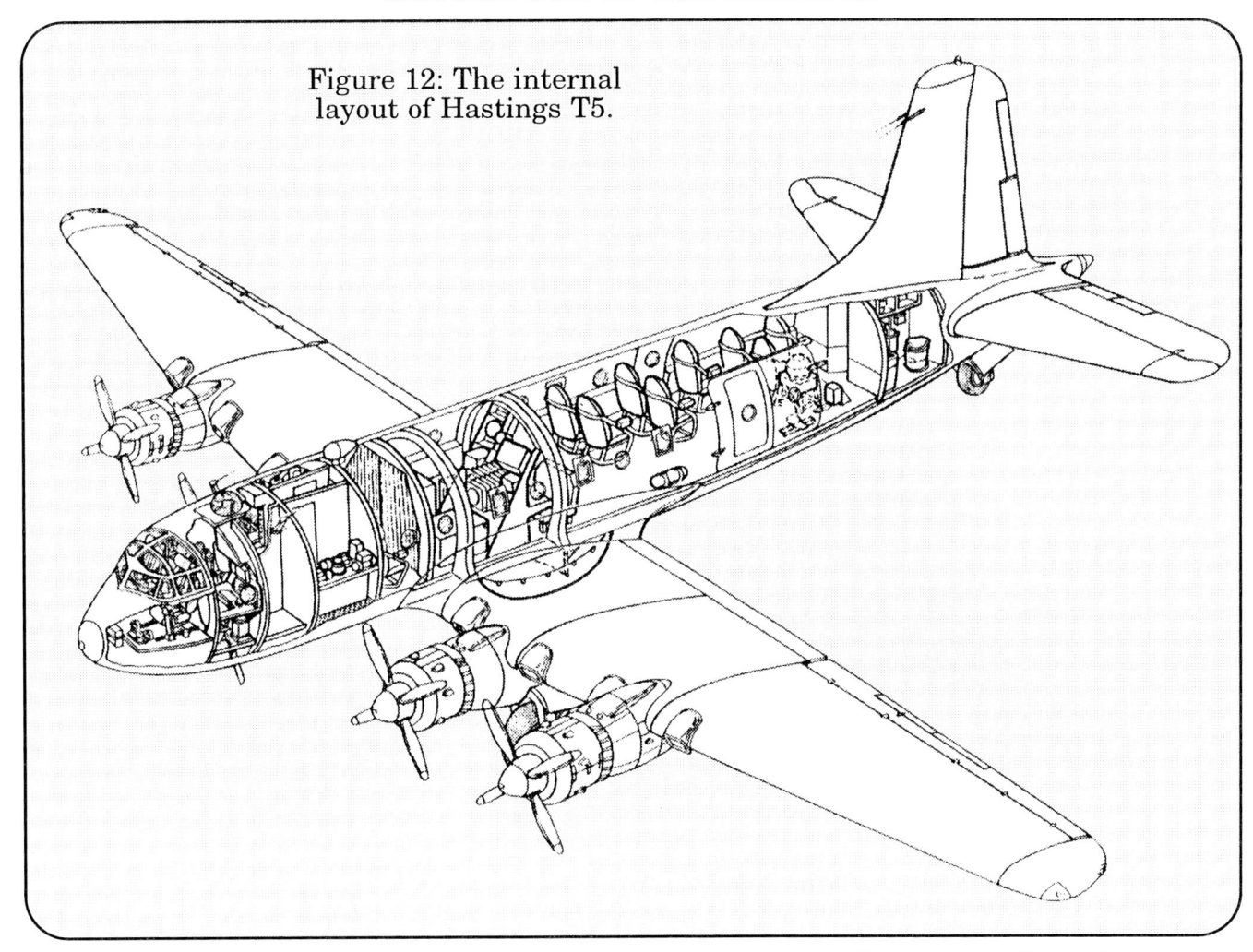

Figure 12: The internal layout of Hastings T5.

"My second flight was in TG587 on the 29th (June) and this was a bit more exciting, as I had to stand behind the student pilot while his performed circuits and bumps - even now I can remember those bumps! The first pilot's intercom had been giving trouble for some time and I had to do whatever I could up front - while the staff pilot let the student fly the aircraft. The student really bounced the old Hastings on the runway and I hit my head on the roof during the landing. However, I did managed to trace the fault to a loose wire in a terminal block - and the student managed to eventually land the aircraft in one piece".

TG610 was lost at Thorney Island on 17th December 1963 when it crashed on landing and careered through the Radio Servicing Flight building, killing the SNCO there, Ted Lewington, and injuring a Chief Technician.

Hastings courses at Thorney Island would often go overseas to complete their night-flying phrase, to places like Cyprus or Idris in North Africa. Whether this was to minimise the number of complaints about low flying and noise in the local U.K. area or to give the crews night flying experience from strange airfields is not known - probably a combination of both!

A minimum of groundcrew were sent along to give support, both day and night. In one case Junior Technician Paddy Porter in 1962 on the last aircraft (TG533) of the detatchment to leave Idris (Castel Benito to old timers!) had a bit of excitement just after the aircraft got airborne:

"There was a dreadful noise from one of the engines, the Hastings unlike the Beverley did not have a passenger address system. I glanced sideways to look out of the window and nearly **** myself, as I saw a trail of what I thought was smoke streaming past. All it was, I later found out, was fuel being dumped by the flight engineer to lighten the aircraft, so that we would not have to make an overweight landing on three engines."

241 OCU was another unit that operated from Dishforth and also operated Hastings, amongst which were TG55S, TG563, TG566, TG567,TG570 and TG571, some of which would be modified later to Met Mk.I aircraft, but most would eventually join 242 OCU. An exception to this was TG567, which after conversion to Met Mk.I, and 202 Squadron service, joined A&AEE Boscombe Down in 1965 and was struck off charge in 1966.

Having struck-off charge a number of their Hastings in 1959 242 OCU did the same with more in 1967 with the closing down of their Hasting training operations. The Handley Page representative was guest of honour at the closing down ceremony of the Hastings at Thorney Island, and was on hand at the dinner to hear the Wing Commander, who was to give a short speech, having just landed the last Hastings from Cyprus, say:

"When I touched down here today I said to

HANDLEY PAGE HASTINGS AND HERMES

"...and running in from your right ladies and gentlemen" Hastings TG503 of 1066 Squadron makes a low pass at Scampton. [Sqn. Ldr Jackson]

myself at last you have mastered the brute, then it turned round and bit me!"

How many others have said that and bounced a Hastings in?

A number of Hastings were still operated by various units, but only in ones and twos; 115 Squadron was one of these units and operated from Watton and flew I.R.I.S (International radio installations and systems) checks. This was to check navigation and approach aids for the Inspectorate of Radio Services. The aircraft were named 'Iris II' and 'Iris III', one being WJ338. Maurice Patterson was one of the signallers on these aircraft:

"Our call-sign was 'Iris' and all the Air Traffic Control set-ups knew this, between four to six Group Captains and/or Wing Commanders used to fly down the back monitoring every aid in service, be it beacon, radar approach or merely VDP, -- one spurious transmission and - pow!!"

The Inspectorate's checks were usually integrated with the RAF Flight Checking Force, the

The personnel of 1066 Squadron at RAF Scampton. [Sqn. Ldr Jackson]

SERVICE USE OF THE HASTINGS

TG503 if 1066 Squadron carrying out a patrol of the North Sea oil rigs. [Sqn.Ldr. Jackson]

unit operating the Hastings from January 1967 until January 1969, with WJ338 being struck off charge on 4 July 1969.

Hastings were also used at the No.1 Parachute Training School at Abingdon, who operated TG615, WD486 and WD487. Others were used at various establishments such as the TRE and Transport Command Development Unit. Much experimental work to do with transport development and parachuting of a special nature was carried out with the Hastings, and some to do with the Special Duty Flight and Radar Reconnaissance Flight at Wyton. On these aircraft the TR1154/1155 was replaced by the STR18 amongst other modern radio and radar equipment, and the aircrew specially selected.

Bombing School.
The 23 March 1960 marked the introduction of Hastings T5 aircraft into the Bomber Command Bombing School at Lindholme (later to become Strike Command Bombing School). The first arrival was TG517, an ex-Met.1 of 202 Squadron, the first of ten that had been converted from Mk.1 aircraft by Airwork Ltd at Blackbushe. The unit's role was to train navigators for the 'V' bomber force, plus air experience and familiarisation to navigators who had been selected for the Phantom FGRII and Buccaneer aircraft. The unit that flew the Hastings was the Air Training Squadron of the BCBS, which became the last RAF unit to fly the Hastings, by then known affectionately as '1066 Squadron' for obvious reasons.

Externally there was no identifying difference from the standard Mk 1 aircraft, except for the large H2S radome under the fuselage just forward of the entrance door. Internally the cabin arrangement was redesigned to suit the tutorial role (see T5 sectioned drawing), and there was a rectangular hatch in the floor between beams 590 and 610 to give access to the H2S scanner unit. The T5 was an adaptation of the Mk.1 aircraft with all C3 modifications, but was not adaptable or convertible to other roles. The fuselage accommodated the standard five operating crew members and a cabin for twelve personnel under instruction. The latter cabin was divided into a training and non-training section by a black-out curtain at station 300. Aft of this curtain was the NBS equipment of a main console and associated items, with additional electrical equipment being fitted to supply the NBS system, which was comprised of a navigation and bombing system with a bombing computer. Airborne radar equipment formed part of the IFF/SSR radar system and the power equipment needed for the NBS equipment was accommodated in a compartment opposite to the galley. A pressurisation system of four 750 litre bottles was installed to supply dry pressurised air to certain units of the H2S scanner.

Aircrew considered the T5 a reliable aircraft

HANDLEY PAGE HASTINGS AND HERMES

TG503 again, this time seen off Spurn Head on North Sea patrol. [Sqn.Ldr Jackson].

that did a successful job of training with a reasonable ride, although, in the case of the navigators there was some doubt cast regarding the relationship between the training and the type of aircraft they were to fly on as to the relatively low speed of the Hastings. Picture quality and the comfortable environment of the Hastings bore little resemblance to that of the Phantom and Buccaneer. During the period 1960 to 1972 the Radar Flight of 230 OTU, later known as 1066 Squadron, was based at Lindholme and during that period trained over 1,000 navigators. Then in 1972 the unit moved to RAF Scampton and had as its last commanding officer Squadron Leader Jackson, who left me in no doubt that in his considered experience the Hastings was an outstanding aircraft.

As well as the training role 1066 Squadron were called on to operate their Hastings on oil rig patrols, which had commenced because of the prompting of the MOD by an MP, who had wished to know what precautions were being taken in regard to oil rigs, for it was reported that Soviet vessels had been seen close to the rigs taking photographs. A Nimrod squadron was eventually detailed for the chore, but the Hastings was called on to do the job weeks before the arrival of the Nimrods. In this role the Hastings took-off from Scampton and commenced their patrol from Spurn Head on the Humber and covered virtually

An interesting formation - Hastings T5 TG505 formating with DH Chipmunk - a tricky job! [Sqn Ldr Jackson]

HANDLEY PAGE HASTINGS AND HERMES

Hastings TG517 flying in to Winthorpe airfield for a final touchdown to join the Newark Air Museum. [Sqn. Ldr Jackson]

all the oil rigs then in existance up to Aberdeen.

The last of the so-called 'Cod Wars' with the Icelanders found the Nimrod reconnaisance aircraft unable to mount sorties that were required to support the British naval force and trawlers off Iceland. So almost overnight the T5 Hastings were fitted with a radio altimeter, single side-band radio and other equipment to add a further role to their repertoire. The T5 aircraft mounted Vulcan radar in the H2S radome for their training role, so with the pencil beam narrowed down and sweeping 360 degrees they could detect the Icelandic gun-boats still in the fjords, or discover them as they came out to intercept the British ships. Sometimes the Hastings themselves were intercepted by Icelandic observation Fokker F-27 Friendships, who likewise were carrying out their duty on their country's behalf. For each sortie in the Cod War a yellow codfish was painted below the cockpit side window of the Hastings - a new form of battle honour!. If during the Berlin Airlift sorties had been displayed in this manner, coal emblems would have covered a large proportion of the fuselage!

The Radar Flight aircraft were also used in the support of Strike Command for the transport of urgent spares and supplies to the various stations in the United Kingdom and Western Europe, as well as spares to grounded or damaged aircraft. On 22nd June 1977 the first of 1066 Squadron's last four Hastings was retired; this was TG517 which was flown by Sqn. Ldr. Jackson and crew to Wintorpe airfield for the Newark Air Museum. Then TG503 was cleaned up and re-painted and flown to Gatow Airport, Berlin for permanent display to commemorate the Berlin Airlift.

Sqn. Ldr Jackson and crew handing over TG503 and log-book to the Station C.O. Gataw. [Sqn. Ldr. Jackson]

HANDLEY PAGE HASTINGS AND HERMES

Agreement was recieved from all the controllers in Berlin, including the Russians, for the flight over the Western Sector. The flight and handing over was made by Sqn. Ldr Jackson and crew and recieved a great reception from the Berliners. Since then TG503 has had its radome removed and looks more like it did during the Berlin Airlift. The last to go was TG511, which had been allocated to the RAF Museum, Cosford, and again was flown by Sqn. Ldr 'Jacko' Jackson and crew. This took place on 16 August and was the last flight of an RAF Hastings and for TG511 the termination of 29 years service. Meanwhile TG505 had been delivered to St Athans, where it was dismantled and delivered by road to the SAS at Hereford for ground training.

Squadron Leader Jackson as OC Hastings Flight of 230 OTU was to receive from the AOC-in-C the following signal:

"1. As the Hastings comes to the end of its days with 230 OTU I wish to place on record the excellent service which 1066 Squadron has given. Both to Strike Command and to the Royal Air Force. You have not only fulfilled your primary task at all times, but on those many occasions you have been needed you have also augmented and supported the front line and carried out a multitude of other tasks in a most efficient and praiseworthy manner.

2. In spite of the vintage and maturity of the Hastings your spendid safety and serviceability record is second to none and reflects commendably upon your aircrew and groundcrew alike."

Fortunately for posterity and aviation enthusiasts, examples of the Hastings are maintained at a number of sites, for as well as the aircraft already mentioned a C Mk.lA TG528 was purchased by the Skyfame Museum at Staverton, and upon the unfortunate demise of that museum it was acquired by the Imperial War Museum at Duxford, where it has been renovated and exhibited. Another C Mk.lA has contributed its wings, centre-section and other parts to the Yorkshire Air Museum's Halifax project at Elvington. Prior to that it had been in use at the RAF Fire School at Catterick, Yorkshire. Rework of, and fabrication of various parts of the Halifax fuselage etc has been carried out by the British Aerospace Apprentices School Brough as part of their training scheme and the Halifax has now been completed with the help of volunteers. A VIP Hastings C Mk.3 has been preserved by the RNZAF in New Zealand.

One flight that has not been recorded or given any prominence occurred on 7 May 1976, when 1066 Squadron mounted a flight to RAF Wittering thirty years to the day that the first Hastings got airborne, arriving there within five minutes of the original take off time, to be greeted by Air Traffic Control Staff singing over the R/t *"Happy birthday to you"* - the veteran Hastings had completed 30 years service.

So the Hastings, which had started life as a derivative of the Halifax, was a problem child in its infancy, became a loved, reliable and tractable maid-of-all-work, remembered by her crews with affection, was also remembered as a tough old lady capable of many tasks - infact, a Handley Page aircraft of distinction. I will close with words from Squadron Leader K. R. Jackson MBE, AFC, who flew 5,000 hours on the Hastings - *"a truly great aeroplane made by a fine company"*.

Chapter Seven
Hermes aircraft development

Historical background.
The Hermes emerged, like the Hastings, from the HP64 project, when it was hoped to produce an aircraft that could be used as either a civil or military transport on similar lines to the DC4/C54. Unfortunately, like the Avro Tudor and York, it was done on the basis of a wartime bomber, whose origins were in 1936, and it was not updated with features like a tricycle undercarriage or a modern aerofoil section. Contract 3205/C4c and specification C15/43 were issued to Handley Page for two Halifax aircraft converted to passenger carriage and referring to a civil development of the Halifax Mk III with the introduction of a new fuselage and centre-section, the date was 1 January 1944.

At this stage in planning it was foreseen that the military and civil transport aircraft would have similar airframes, and it would therefore be possible to vary the production rate of either the civil or military aircraft on the production line as the demand determined. What was not foreseen was the dithering effect that the first flight of the Avro Tudor I would have on the Ministry's planning and ordering of the HP67 and 68. A contract had been placed before August 1944 for jigging and tooling for the HP67 and Handley Page were preparing the jigs in their experimental shop for the commencememt of production, so as to allow a minimum of delay in change-over once permission was given.

By the start of December Handley Page had requested cancellation of the HP66 and were proceeding with four prototype of the HP67/68, although only one fully equipped Hermes was to be constructed, the priority was as follows:

(a) flying shell representing HP67/68
(b) first fully equipped Hastings to 3/44/HP
(c) second fully equipped Hastings
(d) fully equipped Hermes.

The Ministry at this time felt that they should press the MAP for a decision on whether there was going to be a production contract for the Hermes. They also felt that the fuselage design of the Hermes was handicapped by its connection with the Halifax conversion through the HP64, that if the fuselage was eight or ten feet longer then an improved layout would result. On 5 January 1945 RDAT had a report on a visit to Handley Page, which also gave a comparison between the Tudor II and the Hermes I why the Tudor II and not the Tudor I it does not explain! The Ministry was still keen on Handley Page extending the fuselage length and felt that this lengthening would give a marked improvement to the Hermes. But RDAT made the point that they could hardly approach Handley Page over the lengthening of the fuselage as the only potential customer was BOAC and they appeared not to be particularly interested. This lack of decision and avoidance of briefing Handley Page on lengthening the fuselage continued into mid-February, when the firm proposed that the max weight should be increased to 75,000 lbs.

As doubts were by now being voiced over Hermes production, the MAP in March 1945 confirmed that although the Hermes was not cancelled, neither were production arrangements being made and the Hermes prototype was to take second priority to the Hastings. It was made clear that production of the Hermes would depend on the flying of the Hermes prototype and Tudor II. This indecision on the future of the Hermes was further emphasised, when on 29 May in a letter to the CRD, the DD/RDA stated that a strong case could not be made for constructing the Hermes prototype as the pressurisation equipment of the Hermes and Tudor II was identical. Further pressure for the cancellation of the Hermes occurred when the Tudor I had its first flight on 11 June, although in an unequipped and unpressurised condition, with the second prototype Tudor I in a pressurised condition due to fly in a few months. Certain parties expressed the hope that the DGCA was expected to support the case for continuing with the Hermes prototype on a comparative basis.

Draft specification 15/43 issue 2 was then raised to cover certain design changes on the

Table 13 RDAT comparison between Tudor II and Hermes I.

	Tudor II	Hermes I
Maximum all up weight	75,000 lb	70,000 lb
Take-off power per engine	1660 hp	1680 hp
Wing loading	52.8 lb/sq feet	49.8 lb/sq feet
Number of passengers	40	34
Volume per passenger	68.6 cu.feet	64.1 cu.feet
Cargo volume	730 cu.feet	730 cu.feet

HANDLEY PAGE HASTINGS AND HERMES

Competitor to the Hermes was the Avro Tudor, BOAC's flagship Mk.1 G-AGRF 'Elizabeth of England being seen here. [Avro]

Hermes I that mainly affected the use of the wider span wing, for it was decided to use the complete wing of the cancelled HP66 prototype to speed up production of the Hermes prototype. Although no series production of the Hermes was contemplated, a lot would depend on the flight test results of the Hermes and Tudor prototypes. However, the Tudor I's flight tests did not prove satisfactory by the time that the Hermes I was completed, for aerodynamic problems were encountered, and a redesign of the fin and rudder was required and a shortening of the main undercarriage.

The Hermes I was completed by November 1945 and the firm anxious to get on with the flight tests, for the Tudor I and Bristol 170 had already flown. Taxying tests of the Hermes I commenced on 1 December with the first flight scheduled to commence the following day. Unfortunately, as explained in Chapter 4, the first flight ended in disaster and the death of the firm's Chief Test Pilot and Test Observer. Priority was now placed on the completion of the second prototype, with the decision made to carry out the flight test from Wittering, where there was a longer runway. The work on the second prototype (G-AGUB) was suspended temporarily due to the inquiry on G-AGSS, but discussions behind the scenes resulted in the elevators being metal covered (this increased their lightness in flight), and the combined balance and trim tab was modified to two halves with each a separate function, and the trim tab cables replaced by dural rods so as to eliminate stretch.

Further to this, an increase in fuselage length was now being proposed and Handley Page requested that the specification be altered and that £500,000 be paid for the two aircraft. The Ministry argued against this as the first prototype had crashed.

On 27 May 1946 agreement was reached on extending G-AGUB by 160 inches, 80 inches added to the forward fuselage and 80 inches aft, this would increase the fuselage length from 82 feet 2 inches to 95 feet 6 inches. The specification was amended from 15/43 issue 2 to 33/46 for the Hermes II, which was the new designation now that the fuselage was lengthened and known in the works as HP74. At the same time it was agreed to lengthen the fuselage at no cost to the Ministry.

At a meeting convened at Cricklewood to which BOAC were invited, the radio and radar requirements of the Hermes II was discussed, while at the same time the Hermes IV (HP81) was proposed. For this it was planned to have the long fuselage of the Hermes II, but with a tricycle undercarriage and Hercules 263 engines. Unfortunately, the mock-up was held up by the priority work on the Hastings test wing, and the lengthening of the Hermes II fuselage with the outer wings being modified to accept Marflex fuel tanks.

Handley Page on 1 July 1946 gave to the Ministry of Supply brief details of the Hermes II and offered it in two versions, a 52 seater sitting four abreast, or a 64 seater sitting five abreast, a range of 3030 miles with a 9,300 lb payload or 1920 miles with 1600 lb payload. A BOAC representative visited Cricklewood on 5 July to give advice on cabin furnishing, and suggested a cabin layout of 40 day passengers with 'Empire' seats at a minimum pitch of 45 inches and ideally 48-51 inches - what price the present day Jumbo jet seating pitch? An astrodome was required, similar to the jettisonable one on the Hastings, a crew rest room was also required but not at the expense of passenger accommodation. Shortly afterwards, a letter between two departments in the Ministry suggested that no detailed furnishing should be incorporated in the specification and

HERMES AIRCRAFT DEVELOPMENT

Table 14 Estimated performance between Hermes I and Hermes II.

	Hermes 1.	Hermes II.
Engine type	Hercules 120	Hercules 120
Wingspan	113 feet	113 feet
Overall length	82 feet 2 inch	95 feet 6 inch
Maximum take-off weight	75,000 lbs	80,000 lbs
Maximum landing weight	70,000 lbs	70,000 lbs
Basic equipped weight	48,567 lbs	51,146 lbs
Seating capacity	34-50 seats	52-64 seats
Maximum fuel capacity	2560 gallons	2560 gallons
Payload with maximum fuel	6,900 lb	9,300 lbs
Maximum payload	18,600 lbs	16,000 lbs
Cruise speed for maximum range	260 mph at 25,000 feet	270 mph at 25,000 feet
Maximum cruising speed	290 mph at 25,000 feet	282 mph at 25,000 feet

that they provide an empty aircraft as "... *that BOAC advice on furnishing is an entirely different thing from the BOAC furnishing requirements*".

The British Government's stated air policy post-war was "Fly British", yet it would appear that by 1946 the message had not got through to BOAC, for they had already placed small orders for Constellations and Stratocruisers. The argument had been advanced that if BOAC were forced to operate their present outdated fleet or the British aircraft on offer, then the British taxpayer would pay a bigger bill - it has been noticeable that American aircraft did not reduce BOAC's deficit, not to mention the dollar earnings needed to buy the aircraft!

The D/RDA on 15 August wrote to Handley Page about the swing on take-off of civil aircraft, commenting that the swing considered acceptable on military aircraft would not be tolerated on civil aircraft. It was also stated that the Ministry were now interested in the Hermes II and requested Handley Page to investigate the possibility of incorporating a tricycle undercarriage on the aircraft, and what delay would be imposed on production if a go-ahead was authorised. Within the month the firm notified the Ministry that incorporation of a nosewheel undercarriage was quite practical. Meantime the Hermes II was progressing through the shop in its new form, although the Hercules 120 allocated to it was an untried version of the basic Hercules 100 series, even though they had been subjected to the ARB engine bench test and strip procedures. Bristol Aero Engines considered that the Hermes II flight trials should be looked at from a development aspect to provide information and experience applicable to the Hercules 263 engines for the Hermes IV.

During September 1946 the possibility was broached of the MCA ordering a fleet of twenty-five Hermes II with a tricycle undercarriage. This hardened into fact in the November when the Civil Aviation Requirements Committee approved a requirement for twenty-five Hermes IV with Hercules 263 engines for BOAC, and Handley Page was invited to draw up aircraft data for further consideration, and quoting delivery dates and price.

The equipment for the new British aircraft types was mainly of new design and relatively untested, and decisions were having to be made. For instance, comparisons were made between the Normalair Roots type blower and the Airesearch centrifugal type, and the latter was considered

Hermes Mk.2 G-AGUB at Radlett, after the fuselage had been lengthened. [N. Brailsford]

HANDLEY PAGE HASTINGS AND HERMES

better as it could deliver a controlled volume of air to the aircraft cabin. Considerable trouble was also being experienced with the control of the Janitrol type heaters. Electro-Hydraulics were having to carry out considerable redesign of the undercarriage for the Hermes to try and meet the ARB requirements for turning and swing, whilst the MCA and BOAC were concerned over the loss of range with a tricycle undercarriage due to the increase in weight.

By 30 November a meeting took place between Handley Page and BOAC at Thames House to discuss the proposed Hermes IV design, but there did not appear to have been a great deal of rapport between the two sides. At the same time a further development of the Hermes II was projected, this was the HP79 to be powered by four Bristol Theseus prop-turbine engines and was designated the Hermes III.

On 8 March 1947 the ARB notified the MOS that they were prepared to issue a Certificate of Airworthiness (CofA) for the Hermes II with Hercules 120 engines, without further testing of the engines, but the overhaul life was restricted to 150 hours. By the middle of June the Hermes II was 85% complete, although delayed by the pressurisation of the Marflex fuel tanks and fuselage. A month later 90% of the construction was completed, fuel tanks pressurised and the fuselage pressurised to 72 psi. The first flight of G-AGUB took place on 2 September, handling appeared satisfactory and two days later ten hours flying had taken place.

At this date a number of structural items did not comply with the current ARB strength requirements, and on 23 September a letter from Handley Page's Chief Designer's office to the ARB referred to these items, one of which referred to the fin web to boom riveting, which was considered as to be below the factor 0.8 for the gust case at 12,000 feet - yet the identical component would be in service with Hastings aircraft without any trouble being experienced. Further flying continue with the Hermes II, including an appearance at the SBAC Flying Display and some stability tests, the latter indicated tab stability throughout the speed range satisfactory except for a small area.

The Hermes structural strength was the subject of correspondence in the Ministry during November 1947, when it was considered that the construction restricted the maximum continuous cruising speed in level flight to 90 percent. It was also felt that the landing reserve factor on turning on the ground was poor and that the company must do better.

On 21 November 1947 the RTO at Handley Page reported to the Ministry that drawings for the redesigned rear fuselage with lowered tailplane for the Hermes II were almost completed, and it was hoped that this would rectify the stick free instability found on the initial flight tests. This was followed on the 19 December with the alterations to the fuselage frames for the lowered tailplane having been completed, and the tailplane completed in the January 1948. This activity resulted in flight trials having recommenced by the end of the month, when it was found that the lowered tailplane had improved the stability, although the aircraft was still unstable with power 'on', flaps 'down' and CG aft. As a temporary measure a bungee was added to the elevator to cure the latter problem, but further flight trials determined that this did not completely provide the cure when G-AGUB was flown with its CG 54 inches aft and engines at rated power.

Further flight continued with trials to check the suitability of the propellers, cabin heating etc, and by the end of February the firm were satisfied that stability with the lowered tailplane was sufficient for the issue of a CofA. The Hercules 763 powerplants were the subject of interest from the

Left to right: Stafford (designer) N. Brailsford (Flt. Observer) McRostie (Works Superintendant) R. Steel (Flt Observer) Sqn. Ldr Hartford (pilot) [Handley Page Assn]

HERMES AIRCRAFT DEVELOPMENT

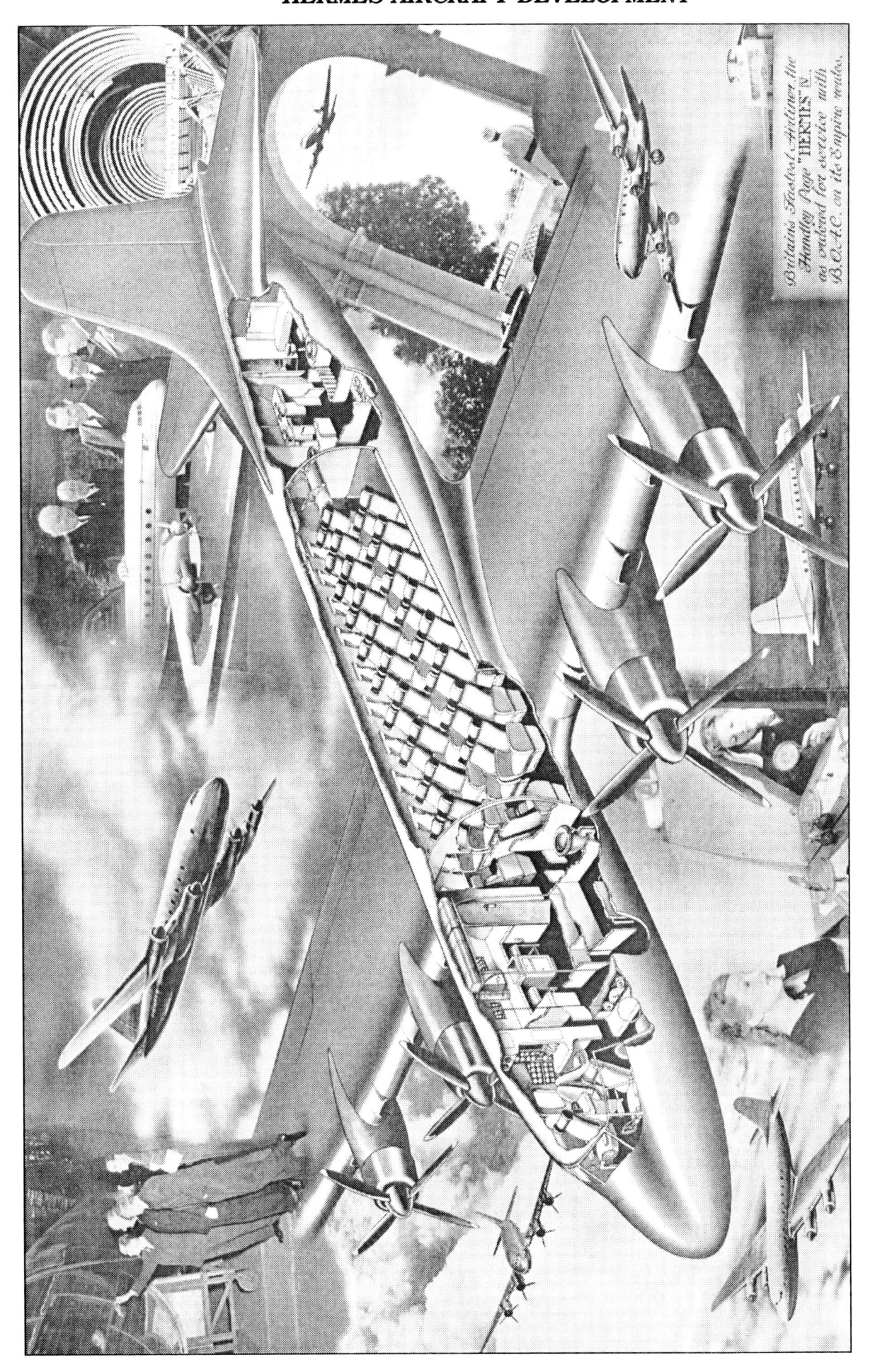

The double page spread taken from a Hermes 4 Sales Brochure that also announced the order from British Overseas Airways Corporation for a number to be used on the Empire Air Routes.

73

HANDLEY PAGE HASTINGS AND HERMES

Ministry, Handley Page and BOAC by this date, for they were anxious to get them fully flight and ground tested, and in temperate and tropical conditions. While the Ministry wanted the tests carried out on the Hermes II, Handley Page suggested a Hastings as Stafford was using the Hermes II for extensive trials with the air conditioning system. G-AGUB was also fitted with water ballast tanks for full load tests and the contractor's trials were completed by 22 May. These trials included dive and low speed tests at full load with the CG aft, and gave the stalling speed with power 'ON' and aircraft clean as 85 knots (98 mph) with a small amount of buffeting. The initial climb to 6,000 feet was accomplished in five minutes.

On 9 July 1948 authority was given for G-AGUB to be transferred to the A&AEE for comparitive stability tests against a Hastings in standard form. From there, on 13 August, G-AGUB returned to Radlett for the fitting of Hercules 763 engines into the outer engine nacelles. The flight trials in this engine configuration commenced in February 1949 and went on until 13 March, when thirteen flying hours had been carried out. The aircraft as it now stood had a lowered tailplane with the incidence at zero, elevator modified with a blunt nose and metal spring bias and the rudder trim ratio altered. Testing was all finished by the end of April, and the aircraft was awaiting collection by BOAC. Going by the records it would appear as if BOAC were in no hurry, for it was almost a month before they collected it for their trials.

The Middle East hot tests of G-AGUB started on 15 August 1949, when it was flown first to Castel Benito (Idris) at a take-off weight of 74,917 lbs. During the flight a performance climb to 20,000 feet was made, as well as a number of level flight checks and range cruise checks at altitude. The flight was without incident and the 1280 miles covered in a time of approximately 52 hours. The following day was concerned with the test programme, which covered take-off, climb, level cruise and engine temperatures, the ground air temperature being 90 F (32 C). As ground temperatures dropped during the following days, permission was sought and received to fly to Khartoum, and on 20 August test flying commenced at Khartoum, where, at the ground level altitude of 1260 feet above sea level the temperature was 102 F (39 C). The BOAC test programme was run through and completed by the 22nd, and the rest of the programme then continued. The Hercules 763 engines operated successfully and few snags occurred, but the need for cabin cooling was very apparent during the trial, when the inside of the cabin reached temperatures of 110-120 F (43-48 C) on the ground. The flight test crew was led by Captain Fry of BOAC with flight test observer E.N. Brailsford representing Handley Page.

G-AGUB returned to Radlett towards the end of March 1950, BOAC having replaced the Hercules 763 engines with Mk.120s. It was then flown with a number of different installations, including magnetometers, research cameras and different engines, until in February 1952 it was registered VX234 and transferred to the MOS Fleet and eventually allotted to RRE Pershore, where it was struck off charge in 1968 and sold as scrap.

Hermes IV.

This was a natural progression from the earlier Hastings/Hermes design and continued with the same basic wing design and aerofoil section. It evolved in early 1946 and by late 1946 had received the approval of the CAR Committee for twenty-five airframes for BOAC to specification 37/46. Two versions were projected, a day/ night version for 18-20 sleepers and a day version seating 36-40 passengers. The tricycle undercarriage was to be produced by Electro-Hydraulics (successor to Messier) and this included a single nose olco with twin nose-wheels. During the year equipment suppliers were contacted over the various systems required by the Hermes IV, which in a number of cases placed the component or system manufacturer at the extremity of its know-how. For while British manufacturers had been forced to concentrate during the war years on conventional military equipment, American manufacturers had through civil and military contracts been able to stride ahead in the development of pressurisation, environmental and electrical equipment. Westland, having developed some pressurisation units for their Welkin highaltitude fighter, were now beholden through their associate company Normalair, to Airesearch of the USA for the license production of pressurisation equipment.

The equipment industries of Britain post-war made great efforts to modernise their products and introduce new technology, which triggered innovative designed equipment for the Hermes airliner. Admittedly, some of the equipment was based on U.S licensing, and much was based on ideas and research which had been done during the latter stages of the 1939-45 War, but which could not be utilised fully then. Basically, Handley Page, Vickers and De Havilland were catalysts to encourage these developments by specifying the needs. Instrument technology was developed, again based on U.S. designs, including the Smiths' electronic SEP.1 autopilot. A new range of radios for communication and navigation were introduced, although some of this came into service first with the RNZAF Hastings. Cross licensing between the USA and the U.K had a long history, both Smiths and Plessey using U.S designed equipment.

The electrical power generating system adopted for the Hermes was a new concept on British aircraft. Alternators on the port and starboard inner engines provided 125 volt AC power to a transformer-rectifier system, so as to provide a high voltage distribution supply. The employment of alternators was to overcome the high altitude brush problem suffered with DC generators, but the rectifiers were to be a source of

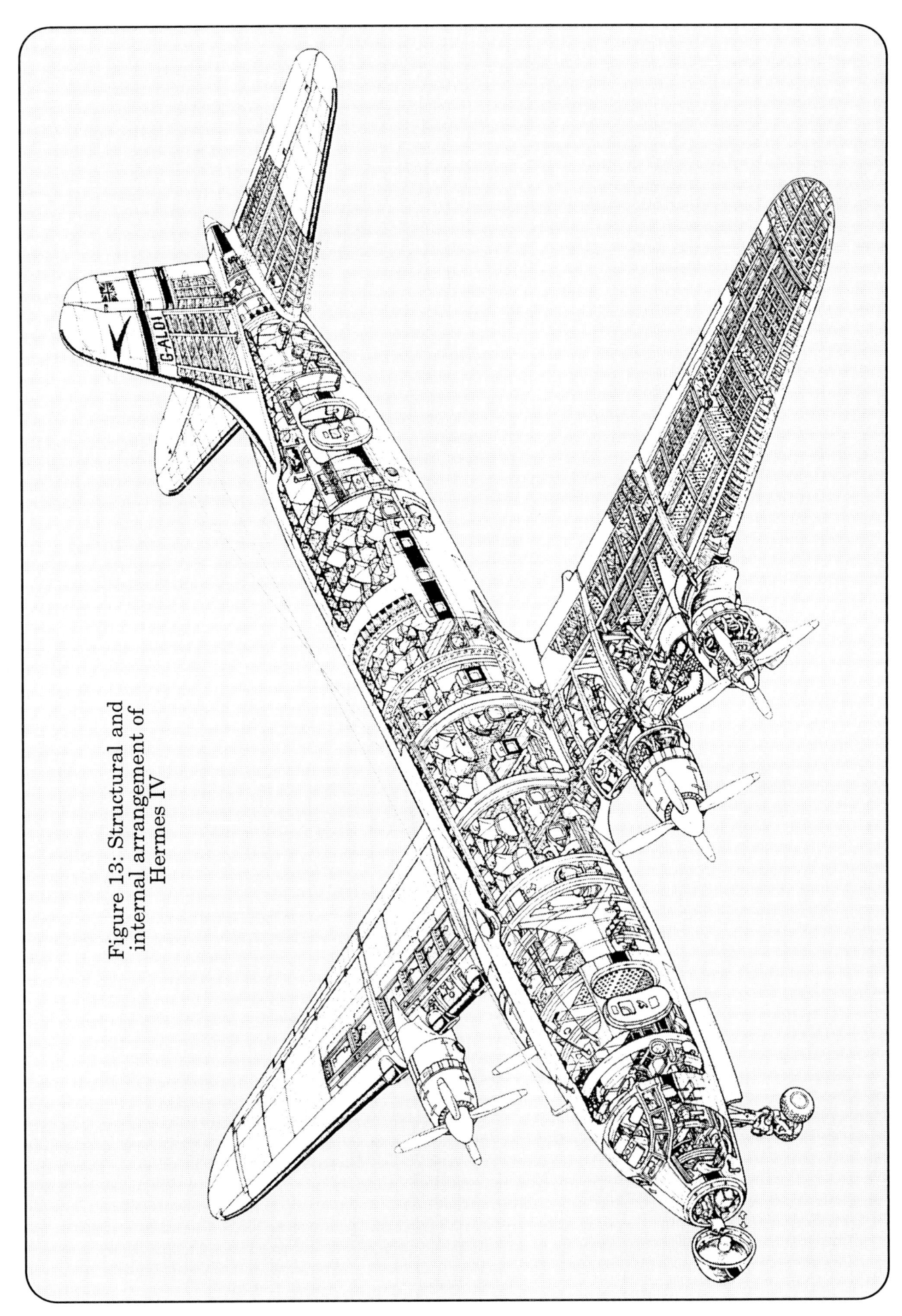

Figure 13: Structural and internal arrangement of Hermes IV

HANDLEY PAGE HASTINGS AND HERMES

unserviceability in service. After the ditching of Hermes IV G-ALDF off Sicily, when both engines became unserviceable, the ARB insisted on a third generator being fitted, this was installed on the starboard outer engine.

Peter Cronbach had responsibility for the Hermes electrical and radio installations, and in correspondence with the author pointed out that some of the electrical requirements stemmed from BOAC's demands, one of which was cooking facilities (not reheating) for meals. The level of power required by the new system needed a higher voltage to reduce cable weights significantly, and as the alternators were running at higher speeds than generators and were inherently lighter, then the power/ weight ratios improved considerably. As he pointed out, solid state power rectifiers were not at that time available and electronic controls were just being developed, so the voltage control for power systems was by carbon pile regulators. In the case of the Hermes the main regulator was stretched to the limits of its capabilities when driven at high engine speeds with low load. For instance night flying at high altitude, when the high resistance required led to the carbon piles being lightly held in contact, resulting in arcing sometimes occurring across the mating faces.

Regarding propeller equipment, Handley Page contacted De Havilland Propellers on 4 December with an enquiry concerning feathering and braking propellers for the Hermes IV. The firm offered two alternatives, one having NACA series 16 section blades and the other one the Clark Y section, although the firm's experience indicated that the Clark Y section blades had lower vibration stresses.

One company not contacted was Flight Refuelling Ltd, but during November 1946 Sir Alan Cobham had written to Handley Page and stressed the advantages that flight refuelling would confer on the Hermes on the North Atlantic route, such as the increase in payload and range. Apart from the technical problems that would face BOAC, chief designer Stafford in a memo to "HP" stated that although he did not dispute the facts in the letter, he was not convinced that on the North Atlantic the reliability of flight refuelling could be maintained under all weather conditions.

Then during October 1946 a visit by Handley Page representatives to the Ministry of Supply gave them food for thought, for the impression was that the Ministry people were not happy with BOAC's attitude regarding the Hermes and Tudor aircraft. As it appeared as if BOAC not only favoured Lockheed's Constellation airliner, but were unconditionally accepting Lockheed's performance figures, yet at the same time interpreting Handley Page weight and performance figures their own way. This intrigued the Handley Page personnel, as they found some of the BOAC calculations contradictory. For instance, northbound from Johannesburg's airfield the payload to Nairobi (a high altitude airfield) for the Constellation was 10,409 lbs and nil for the Hermes. As a matter of interest, the Tudor II had already been assessed by BOAC as having nil payload out of Nairobi, yet in 1952 the author was a crew member of an independently owned Tudor II that carried out a flight from the U.K to Johannesburg via Nairobi, and its payload was 75 passengers southbound and a load of freight and a few passengers northbound.

With the change from Hermes II to Hermes IV structural design changes were required, for the centre-section was then required to cater for the change-over from tailwheel undercarriage to tricycle and the landing loads would be through different points. Further changes were also required to cater for different tanks in the outer wings as well as equipment location in the fuselage. With the result that this change-over and resultant design led to quite an amount of delay in producing the first Hermes IV.

Handley Page were by now also in the design study stage of a further development, which was the final development of the Hermes airframe, this was the Hermes Mk.V, a Bristol Theseus gas turbine powered version of the Hermes IV. It was intended that this version would operate up to altitudes of 30,000 feet. It had a different flap system to the Hermes IV, having high-lift double-slotted type flaps. Maxaret anti-skid brakes were to be installed as well as thermal wing and empennage de-icing. The concept was advanced, but died an early death as the Theseus engine development was curtailed and only two aircraft ordered.

On 25 October 1946 a visit by Electro-Hydraulic's personnel was paid to Cricklewood to discuss the Hermes undercarriage. The result of this was hardly pleasing to Handley Page, for an increase in weight above estimate had taken place, an occurrence that would be repeated with other items of equipment for the Hermes. The weight of each main undercarriage was 470 lbs less wheels, but the nosewheel undercarriage was quoted as 520 lbs against the originally quoted weight of 400 lbs. The increase was claimed to be due to the increased weight of the steering gear. The operating pressure of the Hermes IV hydraulic system had been raised to 3000 psi, and it was estimated that the undercarriage retraction time would be 8 seconds. The same company were developing a disc brake of French design, but eventually the decision was made to install Dunlop wheels and brakes, as neither Palmer nor Goodyear would give performance assurances.

Handley Page staff had been making comparison checks between their Hermes IV and American aircraft, with a more detailed one being made in December 1946 of the Douglas DC6. The method of computation of the direct flying costs was based on the ATA (of USA) formula as used in the DC6 brochure. From this it was calculated that the:maximum payload of the DC6 13,350 lbs (Hermes 13,650 lbs) operating altitude for DC6 19,000 feet (Hermes 25,000 feet) airspeed for the DC6 300 mph (Hermes 271 mph).

The costs of the two aircraft types differed appreciably, partly due to the different pay rates

HERMES AIRCRAFT DEVELOPMENT

The first Hermes IV G-AKFP at the SBAC display at Radlett. [W. Goldsmith]

between the USA and Britain, bare airframe of the DC6 $633,000 (Hermes $254,000) complete airframe of DC6 $733,000 (Hermes $316,400) These were the points that "HP" would try to drive home during his sales talks and correspondence with prospective buyers, unfortunately, by then most were American orientated

The furnishing and passenger layout to suit BOAC resulted on 22 May 1947 with the RTO informing the Air Ministry civil department of BOAC's choice of interior layout. Priority was to be given to 'day only' accommodation using fully adjustable seats at 45 inch pitch, but this layout at the present was not to be the standard for production aircraft. BOAC considered that there was the possibility that towards the end of the production run that a number of aircraft would be used on certain services with sleeping accommodation, and BOAC could not give a firm statement on numbers to that standard at that date!

By 1947 BOAC had suffered a number of astrodome failures on their Constellation 749 aircraft, so this brought forth further correspondence on the Hermes as well as calculations and recommendations regarding design between the RAE and Handley Page. The final result of this was a further change in the Hermes design with the deletion of the astrodome and the installation of a periscope type sextant. An exception to this was Hermes IV G-AKFP, the first production aircraft, which was complete with astrodome when finished and for its trials at Boscombe Down in April 1949.

The Chief Executive of BOAC, Whitney Straight, visited Handley Page on 4 September 1947 to view the seating arrangements that had been provisionally agreed between BOAC and the firm. He decided that the cabin should be split-up to overcome the long 'tubelike' effect, and he also did not want the BOAC colour scheme applied to the Hermes. So the firm agreed to fit a transverse bulkhead, but this was criticised as impracticable. Finally it was agreed to fit one bulkhead and the four seats aft of the bulkhead re-arranged to face to the rear. This bickering over seating arrangements and viewing the mock-up went on over the next few months, finally finishing

The neat Hercules powerplants and undercarriage of Hermes IV G-AKFP. [Author]

77

HANDLEY PAGE HASTINGS AND HERMES

The front fuselage of a Hermes IV slowly takes shape in the jigs at Radlett. [Handly Page Assn].

on 23 April 1948 with a discussion over the mock-up. Stafford asked if the specification and mock-up (photographs and information already submitted to BOAC) could be accepted as representative of the production aircraft, but Campbell-Orde of BOAC would not take the decision, stating that it must rest with Whitney Straight, who would be returning to the U.K on 7 May!

The undercarriage and brakes proved another problem point, for as been said previously, out of the three manufacturers only Dunlop had shown any interest. So it was recommended that Dunlop drum brakes and wheels were fitted to the Hermes IV, and if the Electro-hydraulic development of the French disc brake was satisfactory then these would be fitted on the Hermes Mk.V.

The Mk.IV mock-up acceptance was still awaiting BOAC's final approval on 5 June, yet airframe production was under way, and just under a month later all the major airframe components of the first aircraft were at Radlett ready for assembly, the fuselage had been pressurised, but the firm were waiting delivery of the four Hercules engines. These had been delayed at Bristol as the firm were waiting on the delivery of DH CSUs and braking propellers for use on their test bed. Bristol wrote to the MOS pointing out that delay of this equipment would have an effect on engine delivery, as they were due to commence delivery of two powerplants in the August. The propellers listed for use on the Hercules 763 engines were DH four-blade, fully-feathering, reversible CD60/446/1 type of 13 foot diameter.

At a major meeting on 16 August to examine the Hermes requirements and present status, the question arose of tropical climate conditions. Handley Page answered that they had not been asked to build an aircraft suitable for tropical climates and it had not been called for in the requirements that had been approved, (although it must be pointed out that the environmental and pressurisation system of the Hermes was as good as any in U.S built aircraft). Sir Frederick was also asked by the DGCA if he was in a position to give guaranteed delivery dates, and gave the answer that the dates given were target dates and not guaranteed dates.

The first production aircraft (there was no prototype IV) was GAKFP, and this made its maiden flight on 5 September 1948. The flight was initially delayed due to a defect in the nosewheel during the taxying tests. The problem was that defects with the undercarriage units would be repeated a number of times during the testing period as well as during its early service. By the end of the year problems with the nose steering would become the point of discussion at Thames House, for it had been found that the steering torque of 15,300 lb inch did not provide an effective steering and that a new steering unit was required. Nosewheel shimmy was another problem and coupled nosewheels had been introduced to prevent this, but this created the need for a large steering torque during low-speed manoeuvring. Nosewheel steering and shimmy were only the start of the problems, for as 1949 got under way it was found that the undercarriage forgings supplied to Electro-Hydraulics were faulty, there being low-load failures during strength tests, and cracks were found near flash lines. Further meetings were held to discuss the forgings of DTD.683 and methods to improve the forgings, with ARB requesting that all the forgings were machined all over to reduce the possibility of cracks, that steel links replaced the light alloy ones, and that the nosewheel cylinder was replaced with a stronger one.

With Hermes G-AKFP due to go to A&AEE for special performance tests, discussion took place as who would do the test flying, Stafford wished the firm's test pilot to carry it out to, save time, as they were familiar with the aircraft. A&AEE were not in favour of this as they considered their pilots of sufficient experience to offset their unfamiliarity with the aircraft. It was finally agreed that the firm's chief test pilot should accompany the aircraft, A&AEE pilots would carry out the tests, except for tests not carried out by Handley Page; in which case the firm's test pilot would carry these out and the A&AEE pilots would repeat them.

During the test programme at Boscombe Down, thirteen stalls were carried out with G-AKFP on a four hour flight. During the following inspection on 24 April 1949 it was found that a fatigue failure had occurred in a 22 gauge bulkhead at the rear end of the fuselage under the tailplane. The Director of RAE wrote to the DGCA about this tailplane spar web failure and its commonality with the Tudor IV failure. In both cases the spar web was completely cut away on the aircraft centre-line, which placed reliance on the spar booms for shear transference, augmented in the case of the Hermes by the bulkhead sheeting above and below the spar. Flutter of the Hermes elevator spring tab was experienced, but this was overcome by fitting twin tabs without mass balance. The tests at A&AEE with the Hermes IV were not an unqualified success, but the aircraft was returned to Radlett on 31 May so as to continue with its C of A flying programme and rectification.

A further discussion took place on 7 December at the DGARD's office with a view to bringing the Hermes IV up to the requirement's performance level, for Hercules 763 tests at A&AEE had revealed that the engines were below Bristol's lowest guaranteed figures for take-off, the boost pressure being too low. At that time the engines were giving 1900 hp, but the ARB had approved re-rating them to 2100 hp for take-off. Handley Page technical staff, while generally accepting that the thick wing in the low position created drag problems with the fillets, felt that a further problem was created by the engine manufacturer. For while Handley Page wanted 1100 hp for economical cruise, Bristol Aero Engines wanted to use only 950 hp, so as to extend engine life and reliability. Reduced engine power did of course result in the aircraft flying tail down, a position that would not be on the better part of the lift/drag curve. Certainly, Handley Page would have preferred a higher cruising speed, for they would have then set the wings to the fuselage at an angle of about 4 degrees.

Construction of Hermes tailplanes and elevators in fixed jigs. [Handley Page Assn.]

HANDLEY PAGE HASTINGS AND HERMES

Cover page to a Handley-Page sales brochure for the Hermes IV, showing an artist's impression of the aircraft in flight.

Campbell-Orde of BOAC flew on the second Hermes IV G-ALDA and gave as his opinion that the noise level was not as low as he wished or expected, although the flight deck noise level was exceedingly low, and crew members could converse without use of the intercom. The Ministry were apparently expecting some complaints, especially regarding vibration, for an internal letter stated:

"As we may hear criticism from some quarters of vibration on the Hermes, you will be interested in the following summary of information I have received on the Constellation ....". -This was followed by a cockpit placarding from the Constellation 749 with Curtiss electric three-blade propellers. This indicated that the useful rpm available to the pilot was 1850 and 2050 rpm. The comment was not far wrong, for the Hermes was snagged for engine/airframe vibration, confirmation of this was passed by the RTO de Havilland to the MOS, showing that on the propeller vibration report that maximum amplitude in the fuselage (mainly 1E excitation) occurred at 1500 rpm (below operating range) and at 1700 to 1900 rpm. So the critical vibration rpm at cruising on the Hermes was not so restrictive as on the 749 Constellation.

The next source of possible criticism was the trouble of icing up of the engine cooling fan blades - this was also put to the test and was found not to be serious in practice - what next? Well, at that stage the undercarriage, nosewheel steering and engines were all likely candidates for complaints. Even Handley Page also had complaints, for over two months flight testing had suffered rpm surge and two failures of feathering pump units. Even by 1 June 1951 the ARB had still not approved the braking propellers on the Hermes IV, although a number of tests had been made by BOAC and Handley Page, and the aircraft had entered service in August 1950.

The naming ceremony of *Hannibal*, the flagship of BOAC's Hermes Fleet was carried out on 11 July 1950, and Sir F.Handley Page made a speech, which included the following:

"...we would have wished that the Hermes were in service last year. This would have been had this continued to be the interim type airliner which was at first planned "

"Equipment never before installed had been specially designed for the Hermes. It is the first large British civil aircraft to embody a pressurisation and air conditioning system whose present day requirements call for humidification and refrigeration. Its high voltage alternating current system is the first of its kind to be used in civil aircraft".

A month later the tests were still continuing, G-ALDA for instance carrying out a series of simulated landing runs using reverse pitch on all engines to measure the stresses in the aileron

HERMES AIRCRAFT DEVELOPMENT

Table 15: Comparison of Hermes IV tests and specification.

		Specification	Flight test
Four engine speed at 78,000 lbs	at sea level	238 mph	231 mph
	14,500 feet	291 mph	268 mph at 12,300 feet
	23,800 feet	299 mph	266 mph at 22,300 feet
Rate of climb on four engines at 82,000 lbs		725 fpm at sea level	930 fpm at sea level
		715 fpm at 8,400 feet	910 fpm at 5,000 feet
Service ceiling		22,700 feet	22,000 feet
Rate of climb on three engines at 82,000 lbs		295 fpm at sea level	450 fpm at sea level
		255 fpm at 8,400 feet	450 fpm at 5,000 feet
Service ceiling on three three engines at 82,000 lbs		11,700 feet	16,700 feet
Minimum control speed		116 mph EAS	105 mph EAS
Take-off distance to 50 feet on all engines at 82,000 lbs		4,500 feet	4,290 feet

control circuit caused by the reverse air flow, this confirmed them as not dangerous.

By January 1951 complaints from BOAC were arriving at Handley Page, for although a number of defects had developed during the training and route proving flights, the major defects started to arrive once the Hermes was on regular service. The cabin blowers for instance were changed twenty-two times for various reasons, but eighteen of them were due to damage caused by loose nuts, bolts and pieces of disintegrating pressurisation ducting. Other snags were related to ignition and sleeve failures and the alleged unreliability of engine torquemeters, quoting in one case that one leg of a flight had been flown with a cylinder out of operation as the flight engineer was reluctant to use the torquemeters. The failure of the torquemeters was found to be due to severe pressure surges in its oil system, mechanical wear of the torque-plate stop and pistons, and oil seepage from the torquemeter heads.

By this date the first Hermes V, G-ALEU, had been flying for about two years, unfortunately on 10 April 1951 it suffered an engine fire and made a wheels-up landing at Chilbolton. On examination of the aircraft it was determined that it was damaged beyond repair (DBR), and the wreckage was sold to Airwork Ltd. In the meantime the high-lift slotted flaps of the Mk.V and their retrospective fitment to the Hermes IV was turned down as too expensive. The other Mk.V, G-ALEV, had joined the MOS Fleet, and apart from being loaned to Handley Page on three occasions for demonstrations, spent its life mainly on testing at Boscombe Down. On 20 February 1952 it departed to RAE Farnborough for tests of Maxaret brakes and finished its life on wing fatigue tests. Then in 1958 its fuselage was sent by road to Aviation Traders for a mock-up of a freight door.

The fitting of a freight door on a Hermes and its conversion into a freighter was one of many projects over the years. There was also the HP.91 Hermes VI, a lightened version of the standard aircraft; the HP.92 Hermes VII with Rolls-Royce Griffon engines;, as well as the civilian equivalent of the Hastings VI. None would mature into hardware, and in any case the British national carrier, BOAC, were set on buying American aircraft, as well as the fact that the pure-jet and prop-jet aircraft were by now a reality.

Hermes Mk. V G-ALEU at the SBAC show at Farnborough. [Author]

HANDLEY PAGE HASTINGS AND HERMES
Chapter Eight
Hermes in Service

In 1949 A.G.Knivett of Handley Page had the job of organising the technical training at Radlett for BOAC aircrew and ground crews, remembers clearly that there was an "anti-British manufactured aircraft" feeling amongst the crews, which was very personal. The same engineer was in February 1950 attached to BOAC No.2 Line at Hurn airport for the training and introduction of the Hermes IV into service. The first Hermes IV was delivered in February 1950, followed by a further six - Handley Page had originally promised the first five aircraft by 31 January 1949, and although there had beem six airframes ready, they had been delayed by the alterations requested and by subcontracters being behind schedule with their deliveries, as related previously.

These first seven aircraft were used for crew training and proving flights over the African routes. The training period was fairly intensive in that the seven aircraft completed 1,656 flying hours and made 2,850 landings in five months; the training period continued until August 1950. On each overseas flight a company representative flew with the aircraft to cover the maintenance and provide assistance where necessary. A certain amount of teething problems were experienced as with all new aircraft, but were in general corrected by modification action or procedure. In that period of training one aircraft was crashlanded at Hurn, this was G-ALDF, which landed with its undercarriage retracted on 1 May 1950, the cause considered to be due to pilot error.

The majority of faults affected the Hercules 763 engines, but some did concern the undercarriage. For instance, on 19 July 1950 G-ALDB was on a training flight with Captain Gibson training pilot and Captain Kempthorne and First Officer L.Field under training. Captain Gibson demonstrated stalling under various conditions, and the flaps and undercarriage was lowered and raised a number of times without incident. A feathering check was then demonstrated with the number 4 engine stopped and the propeller feathered. Bill Field now takes up the story.

"I then carried out a Rate One turn to port and then to starboard. Captain Gibson then said he would lower the undercarriage so as to demonstrate the drag effect. He selected undercarriage down by the normal method, and after a short period two green lights and one red light appeared on the undercarriage indicater, indicating that the two mainwheels were down and locked and that the nosewheel was in an unlocked position. Captain Gibson asked the E/O to check the under carriage fuses, the E/O replied that he would but that the hydraulic pressure had fallen to a low figure. Number 4 engine was then unfeathered by the E/O and he reported that the hydraulic pressure was now zero.

"Hurn tower was informed and a low pass the tower requested so that they could assert as to whether the nose-wheel was either in a locked or unlocked position. The tower confirmed that the two main wheels appeared to be locked down but the nosewheel was definitely unlocked. The aircraft was climbed to a safe height and Captain Gibson

Crew training! Captain Hill of BOAC in front of Hermes Mk.IV G-ALDH 'Heracles' during training at Hurn. [J. Knivett]

HERMES IN SERVICE

and myself took it in turns to fly the aircraft at various speeds and attitudes, especially nose-up followed immediately by a nose-down attitude trying to 'fling' the nosewheel forward into a locked position. Meanwhile other members of the crew had cut a hole in the flight deck floor above the nosewheel and were making attempts to push the nosewheel into a down and locked position with a 'pole' made by bolting pieces of angle iron obtained by dismantling the carriage of the desert survival pack. After approximately 3 1/2 hours the attempts were successful and a landing made."

The fault was traced during investigation to fatigue cracking in the cylinder wall of the nosewheel 'down' jack. Instructions were immediately issued from Handley Page for the removal of all jacks for inspection. Before the fault was finally eliminated all aircraft undercarriage jacks were removed for inspection four times! Also incorporated was a hole and cover in the flight deck floor for future emergencies!

Bill Field went on with his career with BOAC to become a Captain on Boeing 707s. However, the Hermes still had to get over its 'teething' troubles and one of these occurred during the trials, this was with the undercarriage selector valve, with the warning horn continuing to blow during landing and the landing run. A further fault encountered during this period affected the hydraulic pumps, for failure to incorporate a NRV in each pump circuit meant that upon one pump failing, then the other continued to pump through it.

Nosewheel steering continued to provide a few headaches as did the nosewheel 'shimmy'. BOAC had in their requirements requested a twin-nosewheel to ensure a large 'footprint', which of course required a greater torque to turn the nosewheel unit. To overcome the nosewheel 'shimmy' the twin-nosewheels were coupled through the axle, which initially combatted it, but the 'shimmy' was only finally eliminated by the fitment of centrifugal weights. Unfortunately the coupling of the nosewheels reduced the efficiency of the steering. According to BOAC the steering of the Hermes bore poor comparison with the Argonaut - obviously ignoring the fact that the Argonaut only had one nosewheel, and it was their desire for a larger 'footprint' with a twin nosewheel that originated this requirement.

The introduction of the Hermes down the BOAC African routes soon brought out a number of equipment failures as well as the snag of using 115/145 grade fuel for the engines. The normal fuel at that time was 100/130 octane fuel, so with the Hermes using 115/145 grade fuel it was not only necessary to lay down supplies at the main stations but also at the alternatives. The 115/145 fuel's storage life was roughly three months, so it was fortunate that this time just about coincided with the delivery of fuel to the alternative airfields.

The proving flight and the introduction of the Hermes service is best described by Harry Pusey, who was Line Operations Officer and Planning Superintendent No.2 Line on the introduction of the Hermes:

"In 1948 we introduced the Hermes on the route from the U.K to South Africa along the West Coast routes. In retrospect, the first proving flight was hilariously funny.

"The Hermes had already replaced the York on the services between the UK and West Africa, but was giving very serious problems. So much so that one wag had started calling it the Hermite - "Her might come and her might not, but her's doubtful". So bad did this become that there was a Line Standing Order issued to the effect that anyone calling it the Hermite would be instantly dismissed.

"Anyway, we were at Kano, with a few days in hand, when the service aircraft started falling down all over the place. Rashly we handed our proving aircraft over to maintain the scheduled services, and eventually arrived in Johannesburg, with the then route-trained service slip crew, only a matter of a day or so ahead of the scheduled

The Captain and crew of BOAC Hermes 'Hengist' on the innaugural United Kingdom to West Africa service - 6th August 1950.

HANDLEY PAGE HASTINGS AND HERMES

service commencement.

"But things were not right. The Hermes was the only aircraft on the whole continent of Africa to use 115/145 fuel, so we had to keep to the route - and had positioned emergency stocks only at our selected alternates.

"There was low cloud at Brazzaville and we could not raise Brazzaville on the radio. We tried Leopoldville, as it was then known, but they had a public holiday. We could not return to Kano and if we diverted to Pointe Nor we would use up the emergency fuel stock there, and it would take two months to replenish. So we persisted with Brazzaville, where we got down easily in the event.

"In my best schoolboy French I protested to the chief radio man. In his best schoolboy English he replied to the effect that early on in his period of watch he had a pressing call of nature, so he had left the radio room. Unfortunately, the door had closed behind him and he did not have a key. So he had sent a boy to town to get the spare key from the chef du station, and in the meantime had sat on the floor outside the 'cabine' listening to us calling, calling, calling. When I asked, quite reasonably I thought, why he had not broken in, he was horrified. This was a new airport he said, not yet officially open, and he would not dare to do such a thing.

"The effect on the East African route was more pronounced as the programme slipped further - awaiting a sensible solution of the mechanical problems (mainly micro-switch difficulties) consternation and confusion increased in East Africa, and the Governors and Colonial Office got progressively upset. So in the end, Ernest (Hessey, Manager No.2 Line) sent his famous signal telling everyone the programme was delayed indefinitely and that he was writing in full. Unfortunately the signal was subjected to the usual telegraphese of the day and was received in Nairobi as "Hermes delayed indefinitely now stop writing" instead of "Hermes programme now delayed indefinite (stop) Writing".

On 17 December 1951 BOAC Hermes G-ALDH had to carry out an emergency landing at Lyon, as the No.4 engine threw its propeller and reduction gear, and in its departure has struck the No.3 engine. The landing was faultlessly carried out and no damage had occurred to the airframe and there were no casualties. Then on 28 February 1952 another Hermes decided to throw a propeller and reduction gear, this was BOAC G-ALDI, which forcelanded at Tripoli aerodrome (North Africa) after the No.3 engine and reduction gear flew off, and in departing it had smashed into the starboard side of the fuselage just after take-off. A hole was torn in the fuselage and the first officer slightly hurt.

This was followed on 26 May, when due to a navigational error, BOAC G-ALDN ran out of fuel and crashlanded in the Sahara Desert. The landing was carried out with skill and without casualties, part of which can be attributed to the robustness of the Hermes construction. Unfortunately First Officer Haslam died of heat exhaustion during the wait for rescue. The captain and navigator of the aircraft were instantly dismissed from the Corporation.

Brian Hutchinson was the BOAC Techical Officer (Development Engineer) responsible for the Hercules engines and powerplant, that covered the investigation and rectification of defects, life development, modifications and operational problems:-

"I recall the reduction gear problem. In fact I went out to investigate two such failures, one near Tripoli and one near Lyon, where the propeller and part of the reduction gear literally came off. It was due to a failure of the bearing/ housing of one of the bevel gears, which then moved outwards and trepanned through the reduction gear housing."

This was confirmed by a detailed investigation of the reduction gear carried out by representatives of the Bristol Aeroplane Co, Ransome & Marles (bearings) and BOAC, when it was considered that the failures were the result of

Hermes IV G-ALDI on the ground at Tripoli after ithe loss of it's number 3 engine and propeller in flight. [J. Knivett]

HERMES IN SERVICE

an accumulation of tolerances in the bevel pinion thrust bearing, which had allowed the balls to run high on the inner track to cause high local track loading. The high local loading in turn caused pitting of the inner race and debris jammed the rotating balls, causing the outer track to burst and allowing the bevel pinion to move outwards against the reduction gear case.

In the early stages of operation, not only were the pressurisation doorseals perishing rather quickly, but the fuselage floor of the cabin in the region of the walk-way, the floor beams and longitudinal stiffeners were failing in numerous instances due to rivets failing and tearing the beams. This latter was rectified by local stiffening, and replacement seals of a different rubber material improved the life of the door seals.

Later on with the conversion to high-density seating, further strengthening of the floor structure would take place. Bob Stanbridge was involved in this:-

"The high-density seating modification was carried out by BOAC in the major check No.8 hanger. The seats were attached by pip-pins through holes in the floor and this area had top-hat sections underneath. The modification involved fitting a top-hat section upside down to the existing one.

The various technical delays were upsetting the services how- ever, which was a point brought up by Whitney Straight at a Liaison Meeting at Cricklewood:-

"That by reason of its passenger comfort the Hermes was obtaining its fair share of business on the African routes, but this was endangered by technical delays".

This was borne out by records covering the period January 1951 to August 1953 when there were 2,198 technical delays. Some of the defects trivial and some in the technical log hilarious, such as *"Pilot's cock too stiff",* so the captain then had to be found for a clearer explanation.

While the Hermes was proving more comfortable and quieter than the Argonaut, hence the greater passenger appeal on the African routes, it was proving less economical than it should have been, for the vibration rpm range of the Hercules 763 engines was in the economical cruise rpm range, which meant selecting a higher rpm and increased fuel consumption and shorter range, or a lower RPM and the aircraft flying at a taildown angle with more drag.

Crew opinions differed regarding the Hermes, for instance Captain A.R.F.Thompson DFC, who flew on the Hermes Fleet for two years wrote:

"It really wasn't much of an aircraft. It was slow, didn't really like leaving the ground, especially at high altitude airfields like Nairobi and dragged its tail like a dog with worms on take-off. The initial trouble was with the engines.

I had no less than three engine failures, all at take-off, on a trip to Lagos and back. I do know that the only thing in my mind, which made the Hermes better than the York was the lack of noise. It didn't handle well, it was sluggish and under-powered."

However, in contradiction to this is the opinion of Captain John Cope of BOAC, who wrote:

"I do remember that the Hermes IV (BOAC) was welcomed warmly by the crews at that time (1950) as it was the first of the new Brabazon Committee aircraft to go into service. It was the first post-war one with nosewheel steering, toe brakes and pressurisation. There were snags at first and the regularity wasn't brilliant. The aircraft though, which operated on the African routes for BOAC (No.2 Line) were greatly liked and fondly remembered by all crews who flew her. It is true that they were underpowered and it was hoped that new engines would be fitted, but I'm afraid progress was too fast and all were overtaken by other aircraft at that time."

As well as the engine faults there were initially defects with the de-icing, and several accessory gearboxes failed due to loss of oil and wear on the drives. The 125 volt electrical system would prove a problem as well as the aircraft got older and equipment manufacturers decided on 115 volt systems, for it meant that it was uneconomical to produce 125 volt components, so a number of components, such as the pitot head and fuel gauges, were replaced with 28 volt alternatives.

Two accidents would occur at Kano, West Africa, over about 18 months, the first in April 1951 when during take-off a port inner tyre burst and made it impossible to raise or lock down the undercarriage. Fuel was jettisoned to reduce the landing weight to 72,500 lbs. The pilot brought the Hermes into a successful touchdown and covered about 1000 yards on the runway and the speed reduced to about 10-15 mph when the port undercarriage collapsed and the aircraft skated for a further 20 yards. There were no serious casualties and the aircraft was repaired. Then on 14 September 1952 G-ALDX when landing, overshot the runway into the rough overshoot area and the nosewheel collapsed.

The investigation into the undercarriage failure on G-ALDU failed to determine any defects in the material of the failed parts. It was stated that the port oleo's material was within specification, there was indications that the front lugs had pulled off and the front link bolt had sheared before the rear lower link had fractured. The transverse tensile properties of the main undercarriage forgings were satisfactory as was the tensile strength of the fractured lever rear link. As will be noticed over the accidents and incidents to the Hermes, the integrity of the structure was superior to many other well-known aircraft, and compared to other aircraft the Hermes was never grounded by the ARB while in service with BOAC.

This reporting of incidents and accidents to the Hermes is as a record and not to indicate that such failures and defects were the perogative of the Hermes. For one can trace a similar trail with most aircraft, especially the ones created after the war to new parameters of design and operation, whether they were Boeing, de Havilland,

HANDLEY PAGE HASTINGS AND HERMES

Lockheed, Handley Page or Vickers. So the reading of these problems must be taken in that context. Some problems were hair-raising as the cause was unknown, as related by radio operator Bob Bullock on a flight piloted by Captain Hill on the 27 January 1951.

"We had unaccountable engine failures when about 50 miles south of Lyon. We were carrying two engineers and due to temperatures and pressures going off the clock they feathered one of the inboard engines, immediately followed by the other engine under similar circumstances. We were now on the two outboard engines only, when one of those recorded the same symptoms. Not wanting to fly on one engine we tried to restart one of the inboards, but it overspeeded and was fortunately brought under control and feathered again.

"During this time, with both alternators out I radioed Lyon for an emergency landing and was given assistance. We spent a nail biting 20 minutes just hoping that the outboards would hold out, which they did and we landed OK. To add to the excitement we had the well-known Hollywood comedian Red Skelton and his party aboard, who was heading for the London Palladium to do a weeks show.

"Having landed safely we spent an hilarious night in a Lyon hotel being entertained by Red, who also supplied the champagne. By the way, no faults were found with the engines and it was thought to be due to contaminated fuel."

A number of charter companies, in later years calling themselves 'Independents', were by 1951 successfully running 'Coach' class and tourist class services on scheduled routes, some down through Africa to Accra, Nairobi and Salisbury. This naturally affected the Corporation's load factors and income, as numerous passengers preferred the more leisurely and cheaper flights that nightstopped. To recapture some of this trade the Corporation put in hand a review of some of their aircraft for conversion to high density seating, this included the Hermes. For this Handley Page submitted on the 30 April 1951 a draft specification to BOAC to cover this arrangement for the Hermes IV and listed three schemes:

Scheme A   64 seats comprising 11 twin seats and 14 triple seats
Scheme B   64 seats comprising 14 twin seats and 12 triple seats
Scheme C   69 seats comprising 12 twin seats and 15 triple seats.

Initially the programme was planned to cover conversion of GALDB, G-ALDC, G-ALDG and G-ALDT, but eventually all Hermes were converted.

BOAC would operate the Hermes over a period of about 3-4 years, during which time they were converted to high-density seating, then replaced by the DH Comet jet liner. BOAC, who were in later years jokingly referred to as the 'Boeing Overseas Aircraft Corporation' by the aviation fraternity, appeared to outsiders even then to have a clique of personnel with a pro-American aircraft attitude, with a tendency for some Corporation senior members looking for some reason to buy American aircraft in preference to British. This jaundiced view was even seen in the acceptance of defects in American aircraft, which would have been highlighted if they had been British aircraft. Whilst accepting that the Tudor fiasco and the initial problems with the Hermes prejudiced their futures, it must be remembered that it was the Ministry and BOAC's insistence that changed the initial concept of the Hermes from an interim airliner.

Comments within the Corporation were not all negative and critical, for passenger response to the aircraft's comfort had been freely expressed. So that Sir Miles Thomas, Chairman of BOAC

Hermes IV G-ALDF 'Hadrian' in a sorry state after a crash-landing at Hurn during crew training. The aircraft was repaired, but was later lost at sea off Sicily on 25th February 1952. [J. Knivett]

HERMES IN SERVICE

BOAC Hermes Mk.lv taxi's in on a 'bush' African airfield. [Handley Page Assn]

from 1949 to 1956, writing in the company magazine made the point that the Hermes as first produced was not only the first genuine airliner produced in Britain since the war, it was also the culmination of a number of manufacturers to produce equipment new to the British aircraft industry. He also referred to the teething troubles on the Hermes, but that since its introduction on the African routes it had received good opinions from the passengers.

Sir Miles Thomas on another occasion remarked on the improved results with the Hermes, which was due to hard work borne by the engineering as well as the flying staffs:

"This was evident not only on the regularity charts on the Board-room walls, but in passenger comments on the standard of comfort of the Hermes and in the monthly financial records...."

The cabin staff were to repeatedly hear favourable comments on the standard of comfort of the Hermes, for the standard of comfort set by the Hermes was a byword in airliner environment and furnishing.

One example indicating lack of co-operation was with the engines, the major fault was with the seizure of the sleeves, this was peculiar to the Hercules 763 engines on the Hermes, as only isolated cases post-war had been experienced on other aircraft. Intensive research on test-beds and flight tests on GAKFP at Radlett had produced nothing to account for this. So as a trial installation Bristol increased the clearance between the sleeves and barrels, but BOAC would not agree to a general application, so much time was lost without this intensive testing. To quote Nimmis of Bristol, that any attempt by his company to accelerate modifications of this nature were resisted by the Corporation pending line experience of the trial installation.

To be fair to BOAC, numerous modifications had been made to the engines soon after the receipt of the aircraft, this included the return to Bristol of all reduction gears for the re-cutting of the oil-ways in the trunnion sleeves, crack-checking of the reduction gear bevel gears, cracks in the rear of the propeller shaft and failure of the bevel pinion thrust bearing. Altogether the reduction gears were returned to Bristol three times and the engines twice, the latter was the result of sleeve seizure, invariably caused by local bulging on the side of the common port. A great deal of development and tests were carried out and an increased running clearance of 0.003 inch by machining the barrel bores was introduced. However, with the increased clearance the oil consumption became excessive, necessitating the introduction of another modification, this was the introduction of packs of four thin sharply-chamfered oil control rings.

An early change on the Mk.763 engine was to reduce vibration by the repositioning of the master rods in adjacent cylinders, Nos 6 and 7, instead of the normal opposite ones of Nos 4 and 11. Pistons were also prone to cracking, so the material was changed, but the pistons were in the end life limited to 2250 hours.

In the early days of Hermes operation numerous modifications were introduced by BOAC. One of these was that the rear spar web was found to have cracked at two positions due to high frequency vibration, so had to be reinforced. Then early in 1952 the ARB notified BOAC and the manufacturer that the Hermes with the greatest number of flying hours was to be returned to Radlett for inspection and test. The ARB had placed an arbitary figure of 4,000 hours life on the Hermes wing, and the extension of this life would be dependent on the material tests carried out at RAE Farnborough and the inspection at the works. On 28 April G-ALDJ was returned to Radlett where specimens of the spar booms were cut out and sent to the RAE. This spar work entailed much preparatory work such as the removal of engines, gearboxes, fuel tanks etc - a costly procedure.

The lifing of the wings was actually increased after a preliminary assessment, then the life of 10,000 hours for the intermediate plane and 15,000 hours for the outer plane was laid down, following which the aircraft were to be re-sparred or spar reinforcements incorporated. This work was the commencement of a series of discussions and correspondence that went on for months, both as to the spar life, who was to pay for spar

HANDLEY PAGE HASTINGS AND HERMES

With flags flying Hermes G-ALDS 'Hesperides' throws up clouds of sand during start-up at Khartoum [A. S. Jacklin]

replacement, whose specification for the spar material - this finished with everyone denying responsibility - except that the wings had to be re-sparred! At that date G-AKFP was already stripped down and was having its structure lightened, for this first off Hermes was heavier than the later Hermes. The fuselage and formers for instance had no lightening holes in them, and this was being rectified and the wings re-sparred, then when completed was eventually sold to Airwork.

When BOAC commenced the withdrawal of their first Hermes four were leased to Airwork and their aircrew trained by BOAC, wth some of the company's engineering staff going for on-job training at BOAC and Radlett. As the Hermes leased by Airwork were to be flown on trooping flights, where difficulties could be experienced in some areas in obtaining 115/145 fuel, it was decided to get Bristol to modify the engines to accept 100/130 grade fuel. The engines were modified in other facets, including single supercharging, and the engines were then designated the Mk.773. A number of minor modifications were incorporated on the airframes by Airwork as necessary for trooping and the aircraft went to Radlett for the fitting of rear facing seats (a War Office requirement) and a number of modifications on the floor structure. With the incorporation of the Mk.773 engines and 100/130 grade fuel the aircraft was designated the Hermes Mk.IVA.

An Airwork flight engineer on the Hermes was George Piper, an ex-RAF Hastings flight engineer:

"The performance of the Hermes was woeful

Table 16. Comparison of Hercules 763 and 773 engines.

	Hercules 763	Hercules 773
Compression ratio	7 to 1	7 to 1
Bore & stroke	5.75 x 6.0 inch	5.75 x 6.0 inch
Swept volume	2360 cu.inch	2360 cu.inch
Supercharger type	two-speed	single-speed
Supercharger gear ratio	MS 6.64 to 1	6.64 to 1
	FS 8.30 to 1	
Fuel grade	115/145	100/130
Carburation	Hobson-RAE BI/BH13	Hobson-RAE BI/BHll with water-methanol
Power at take-off	2100 hp at 2900 rpm +15 1/2 psi	2045 hp at 2900 rpm plus W/M
Rated power	1220 hp at 2500 rpm at 12,500 ft	1645 hp at 2500 rpm at 3,000 ft

HERMES IN SERVICE

throughout the envelope, although to be fair, this was in part due to the trooping configuration.

"A trap for a flight engineer converting from the Hastings to the Hermes (I doubt if there were many of us) was the panel layout. Although the engineer faced aft in both types, the Hastings engine controls and instruments read 4-3-2-1, whereas on the Hermes a fundamental re-think led to a 1-2-3-4 layout. Yes, there were many drawbacks to the Hermes, but nevertheless, it must be stated that by about 1955, when the Airwork-Skyways operation had become established, the aircraft plodded dutifully backwards and forwards between London and Singapore with a punctuality and reliability that other operators of more modern fleets on that demanding route were unable to achieve!"

By the time that Airwork, Britavia and Skyways had leased or purchased the Hermes from BOAC, the Corporation's development and maintenance staffs had worked many of the 'bugs' out of the airframe and engines, and modifications from BOAC and the manufacturers had improved the airframe, engines and equipment to achieve a fairly good reliability. However, having leased four Hermes to Airwork, BOAC were considering reducing the number of Hermes operations and the number of aircraft. With a view to this on 1 January 1953 a special meeting took place between BOAC employer and employee representatives to discuss this programme and the future policy of the Hermes Fleet.

This had come about due to the employee's representative on 22 December 1952 being notified of the Corporation's intention to reduce the fleet size. It was then made clear that the Corporation only wanted to operate nine Hermes. The Union representative wanted to know what was too happen with the other ten aircraft and why had the Corporation gone to the expense of converting all the Hermes aircraft to high-density seating, when even the Union side of the Panel knew there would be a surplus of Hermes aircraft in 1953? The Corporation's representative came up with a number of excuses, including the following:

"The limiting factor is the sector length. Over 1,000 miles it reduces its payload to such an extent that it is uneconomical, and the Hermes has always been an uneconomical aircraft".

This statement was disputed by the Union side, who reminded those present that only a few months previously they had been shown figures and statistics at a Hermes meeting to show how well the aircraft was performing. This was indisputable, for on 14 May a letter and chart had been issued covering mechanical and other delays on the Hermes service, and the wording clearly stated that the Hermes, although an untried aircraft had at that date reached an operating efficiency which reflected high credit on all concerned.

The Corporation administrator's knowledge of what was going on within the Corporation appeared at times to be a little adrift, for in this case the Argonaut Fleet as well as the Hermes were being converted for tourist traffic, also part of the Constellation Fleet; whilst passenger preference for the Hermes over the Argonaut had been demonstrated. Further to this, two more Comstellation 049s were to be purchased and six Yorks retained for training! What resemblance a tailwheel York had to modern tricycle-undercarriaged aircraft for training purposes it would be hard to judge.

In January 1953 it was stated in *The Aeroplane* magazine that BOAC had a number of Hermes airliners for sale, with the offer of appropriate spares provisioning, including engines and propellers. Shortly afterwards it was announced that the asking price was £198,000 per aircraft, this seemed a fair price for an aircraft that was only about three years old and had spar modifications carried out at a cost of £45,000 per aircraft. By the July, the Trade Union side at the Local Joint Panel Committee had accepted the withdrawal of the whole Hermes Fleet, stated to be due to the disappointing financial results of the Corporation. What was surprising was their assertion that there was little market for this type of aircraft, as well as the qualification to this of "economic unsuitability"!

Obviously the Independents did not agree with this assessment, and neither did other purchasers over the years. Airwork's leased Hermes from BOAC and modified as the Hermes IVA with re-rated engines had carried out tropical trials at Entebbe and Khartoum to satisfy the ARB. By 1954 Airwork had a two-and-a-half year War Office trooping contract to carry 7,000 troops each year in each direction between the United Kingdom and Singapore. Britavia had put in to purchase six of the cocooned Hermes from BOAC, but the sale agreement caused a seven week dispute at BOAC with the Union, whose members refused to de-cocoon the aircraft unless Britavia paid their engineering staff the same rate as BOAC and the Hercules engines from the Hermes were routed to BOAC Treforest Engine Overhaul unit for overhaul. Britavia issued a write against BOAC as the firm's trooping contract was threatened.

Shortly afterwards the Corporation were in trouble again due to the DH Comet I crashes, and the Corporation found itself short of aircraft. So Sir Miles Thomas made the decision to use four of the Hermes on the East African routes. Meanwhile the dispute within BOAC was settled and one gets the impression that the main cause of all the Union's arguments was 'jobs for the boys', as the remaining Hermes were sold to Britavia and Skyways (see Appendix 4. Hermes histories). Both of these companies signed trooping contracts with the War Office, with Skyways flying eight 68 passenger seat Hermes to Singapore, while Britavia operated six Hermes on trooping to Cyprus, Canal Zone, Kenya and Singapore. The chairman of the latter company commenting that the aircraft utilisation was extremely high, approaching 3,000 hours per aircraft per year, and reliability and economy had exceeded expectations.

Table 16. Comparison of some airliners in 1948

	Douglas DC6	749 Constellation	Tudor I	Tudor II	Hermes IV.
Wingspan	117ft	123ft	120ft	120ft	113ft
Wing area	1,463sqft	1,650sqft	1,421sqft	1,421sqft	1,408sqft
Length	100ft	95ft 1 1/4in	93ft 11in	105ft 7in	96ft 10in
Basic equipped weight	53,623lb	58,970lb	47,970lb	53,318lb	55,350lb
Max all up weight	93,200lb	107,000lb	70,000lb	80,000 lb	86,000lb
Payload		49,280 lb	6,262 lb	13,985 lb	17,000 lb
Max no of passengers	68	64	24 *	60	74
Max speed	355mph	358mph	260mph	295mph	355mph
Max WM cruise	276mph	327mph	230mph	235mph	276mph
Service ceiling	29,000ft	over 25,000ft	26,000ft	25,500ft	24,500ft
Maximum range ++	2,811 mile	5,500 mile	3,360 mile	1,760 mile	3,080 mile
Take-off dist to 50ft	2,196yds			1,500yds	1,362yds
Initial rate of climb	900ft/min	1,280ft/min	700ft/min	740ft/min	975ft/min
Engines	Double Wasp	Cyclone GR3350	Merlin 620	Merlin 626	Hercules 763
Total engine power	8,400hp	10,000hp	7,080hp	7,080hp	8,400hp
Total fuel (imperial)	3,905 galls	3,537 galls	3,300 galls	3,300 galls	3,172 gall

* 12 night passengers. ++ absolute maximum range.

Airwork as the first Independent company to operate the Hermes was also the first to suffer a loss. On 23 July 1952 Hermes GALDB had taken off on a trooping flight with 68 passengers and 8 crew to Malta. Flying over France in the dark everything appeared normal, then just after Paris the starboard outer engine 'cut' twice and then 'ran away'. The reduction gear drive from the engine had sheared and in seconds the gear had disintegrated, and in doing so it hit the starboard inner engine, twisted and unbalanced the inner propeller and caused the starboard inner engine to partly break away from its mounting. The drag on the starboard side and the loss of two engines on that side caused the aircraft to sink at an alarming rate. The aircraft was shaking badly and the control columns were attempting to slam backwards and forwards against the pilot's efforts to keep them steady, with both pilots trying to hold the aircraft from going into an uncontrollable spin.

The remainder of the crew meanwhile had gone about their emergency drills, putting out a 'Mayday' call, de-pressurised the cabin and jettisoned the fuel. The rate of descent had by then increased to 1200 feet per minute and the pilot, Captain Lovelock, was faced with a crashlanding in the dark with no visual references and over unfamiliar ground. As the night was pitch-black, the few isolated lights on the ground indicated that they were over open country. Lovelock eased the aircraft round in descent to get her into wind for touchdown, then at approximately 700 feet he put the landing lights on, which revealed open farmland as well as electricity pylons. Flattening the descent the pilot eased the Hermes to a crashlanding, sliding it under power cables and avoiding the pylons the aircraft crashed down at about 125 knots.

Slithering along for approximately 100 yards, the hanging starboard inner engine dug in and the aircraft swung around and came to a halt. The crew and passengers evacuated the aircraft, which due to fractured fuel lines and in spite of fire extinquishers operating, burst into flames. The passengers, troops of the Border Regiment, were fallen in by their NCO and a roll call taken. All passengers and crew had escaped injury except Captain Lovelock, who had a damaged back. Without doubt this escape was due to the rearward facing passenger seats, the strength of the Hermes and the supreme airmanship of the crew, and in particular that of Captain Lovelock.

The same company a month later were to lose a second Hermes, this was G-ALDF, which on 25 August was on a flight from Blackbushe to Khartoum via Malta. Whilst over the Mediterranean Sea its starboard inner engine 'ran away', followed by failure of the port inner engine. Unable to control the 'run away' engine it was shut down and the propeller feathered. With the aircraft on two engines the captain was forced to ditch the Hermes in the dark off Trapani, Sicily. Although fishing boats eventually rescued the passengers, seven lost their lives in the meantime.

Bernard Wheeley was an engineer-inspector with Airwork during the period and says of the Hermes:

"When we received the Hermes I had the job of checking with Bristol the mod' state of the engines against the logbooks, as there appeared to be some discrepancies.

"We liked the Hermes, it was easy to work on, there was not a lot of trouble, although it had its particular regular snags like any other aircraft. The two that come to mind are the gearbox bolts, which used to shear due to rpm frequency; the other was the transformer rectifier. As an airframe and engine

HERMES IN SERVICE

Hermes Mk.IV G-ALDA, the former 'Hercuba' of BOAC is seen in full Airwork colours during 1956. [Airwork]

inspector I really enjoyed working on the Hermes, it was quite reliable in general and they were not often in the hangar due to unserviceability.

"After the second ditching we had to fit a generator to an outer engine, making a total of three, to provide electrical power should one of the inner engines have to be shut down. The engines and accessories were very accessible as all the engine cowlings were hinged out of the way. I believe that the life of the Hermes was limited by the main spar life, which I think was 30,000 hours flying time."

During a training session on the Douglas DC6B the author was discussing the Hermes with Dave Haslum, an Airwork flight engineer, and it was obvious that he was more than satisfied with the Hermes aircraft from an operating point of view, but disappointed that the engine vibration range made it impossible for the aircraft to show its true potential. The aircraft being flown at the lower end of the rpm range, with the result that the Hermes flew in a tail-down position and thus not on the best part of the lift/drag curve. However, both of us agreed that the Hermes flight deck was 100 percent better equipped and laid out than the Douglas DC6.

By late 1955 the Hermes had covered in passenger service more than one thousand million passenger miles and carried nearly a million passengers. Hermes G-ALDU of Britavia had achieved the honour of being first of its type across the Atlantic to the USA, when it flew from

Seen on a wet apron at Blackbushe in military guise whilst engaged on Ministry of Defence trooping flights is this unidentified Hermes IV of Airwork. [B. Wheeley]

HANDLEY PAGE HASTINGS AND HERMES

Blackbushe to New York carrying a relief ship's crew of 39 seamen for a New Zealand merchant vessel. The Hermes operated at a maximum all-up-weight of 86,000 lbs out of Blackbushe and flew via Shannon and Gander. The return flight of 39 seamen was of 11 hours duration from Gander direct to Blackbushe.

Amongst the million of miles flown by the Hermes aircraft in service very little of interest was reported, as they were safe passenger miles, so received no headlines. One little-known incident during the trooping flights is of interest from a different angle, it occurred on an Airwork Hermes trooping flight to the Far East. On approach to land the nose undercarriage failed to come down, irrespective of the emergency action and crew prodding with the 'pole'. So the aircraft's captain asked for volunteers from the troops on board to position themselves aft to change the trim, so that he could carry out a mainwheels only landing. This was carried out and a successful landing carried out, the nose only dropping when the aircraft had more or less stopped. The damage was so slight that the aircraft was jacked up and the nosewheel freed, temporary work carried out and a retraction check made, then the Hermes was ferried back empty to its Blackbushe base for full inspection and repair - arriving back within three days of the incident. Clearly indicating, not only initiative on the captain's part, but a tough airframe as well as an aircraft not too hard to fly on its aft C of G.

A magazine correspondent writing about the Hermes and its reliability with an Independent company stated:

"Crews like the Hermes for that reason and for its quietness and comfortable cockpit layout. Indeed, the pilot's seats are universally admired as the best seats ever built. The strength of the aircraft structure is always a comforting thought at the back of their minds".

The author's own experience tends to confirm the general reliability of the Hermes airframe, but its Hercules 773 engines for some unknown reasons never achieved the long-life and reliability of other Hercules engine installations. As the Mk.773 engines neared the end of their overhaul life the oil consumption had increased to the point that it limited the flight stage length. However, it must be admitted that by 1955 the majority of the teething problems were over and the Hermes was earning its keep. To be fair to BOAC crews who operated the Hermes IV, on its introduction into service it was a little bit more sophisticated than some of their previous aircraft, as it had a tricycle undercarriage, full pressurisation and humidifier systems as well as an alternating current electrical generating system. With regard to operating the Hermes, it must be pointed out that the aircrew of the Independent companies often operated away from scheduled routes and company maintenance personnel, often into second-rate airfields, so became more acquainted with their Hermes aircraft than BOAC aircrew were able to do.

This type of experience was illustrated in the Handley Page service engineer's reports, highlighting that the lack of familiarity with the aircraft's systems was often the cause for complaints. For instance, referring to the hydraulics, the reports tend to confirm that inexperienced crews were more likely to use the emergency system when a service was not operating or false indications occurred. An unreliable feature of the undercarriage system was the inconsistent operation of the 'down-lock' indicating system. Whilst the experienced crews carried out a reselection to overcome this problem, the less experienced crews tended to use the

Replacing the fin during a C. of A. overhaul on Hermes Iv G-AKFP of Airwork Ltd at Blackbushe.
[B. Wheeley]

HERMES IN SERVICE

The excellent means of access to the Hercules 773 engine of a Hermes Mk.IV A [B. Wheeley]

emergency system to prove positive 'down-lock' operation - the fault quite often being dirty or bad electrical contacts.

By 1956 the Independents were finding themselves busy in the Middle East as well as trooping to the Far East, as due to a disagreement with the British Government President Nasser of Egypt was expelling British subjects, the Suez Canal was eventually appropriated and this led to the Suez War of 1956. The result was more work for the Hermes, which were averaging 200 hours per aircraft per month with good serviceability and the flying only limited by the availability of aircrew. This intensive flying under arduous conditions, with only basic needs and little sleep at Tripoli Airport (as the local natives were unfriendly and so no outside hotel accommodation was available), placed a heavy load on both the aircrew and groundcrew. This is reflected in the following accident.

Shortly before midnight on 5 November 1956 Hermes G-ALDJ of Britavia crashed while approaching to land at Blackbushe Airport. The aircraft was on charter to the Air Ministry and was evacuating servicemen's families from Tripoli due to the Suez crisis. The visibility at Blackbushe was poor and it was assumed that the captain was letting down on ILS. The aircraft undershot the runway and hit a beech tree 3,617 feet short of the runway, the port wing was damaged by the impact and the aircraft swung sharply to port, finally coming to rest in some fire trees, where it caught fire. The captain and two other crew members were killed on impact, with the flight engineer being the only operating crew member to escape. The aircraft was carrying 74 passengers, and four of these, all children, were lost in the resultant fire. The accident report stated that certain of the mothers ignored cabin crew instructions and put their children through the emergency exits on to the wing, unhappily the fire outside caught these children and they were burnt to death.

The MCA investigation into the accident determined that the aircraft had no technical defects and the undercarriage was down for landing. The flight engineer, who was knocked out in the crash, along with the steward and stewardess acted promptly to get the passengers out of the burning aircraft. No positive conclusion as to the cause of the accident was reached at the Public Inquiry, although it was stated that the crew were tired due to the ten hours spent in unhappy conditions at Tripoli Airport - the author was at Tripoli Airport under these unhappy conditions and can confirm that due to the emergency the chances of sleep was remote.

The reason for enlarging a little on this particular accident is to stress the structural integrity of the Hermes airframe again, for it must be noted that the wing struck the tree, it was damaged but did not break off, and likewise the

HANDLEY PAGE HASTINGS AND HERMES

Hermes V G-ALEU at Radlett during it's flight testing. The view shows the neatness of design of the Bristol Theseus powerplants. The aircraft was damaged beyond repair in a wheels-up landing at Chilbolton. [Handley Page Assn]

fuselage stayed intact in spite of crashing in fir trees, and allowed the evacuation of 73 survivors. In the author's opinion it is doubtful if any type of U.S built aircraft at that time would have stood up to this sort of treatment.

The Hermes aircraft of the British Independent operators in those days operated no scheduled routes, but operated trooping contracts and charters, these to the annoyance of some BOAC Union members and Socialist politicians. Their passengers ranged from H.M troops to distinguished orchestras, there were complete ship's crews for ships abroad, and the transport of goods, this included the transport of rhesus monkeys for U.K and USA medical research. These latter passengers created a problem of their own, for the little creatures' inhibited habits created a distinct smell and quantity that the Hermes environmental control was unable to cope with. To combat this, products such as Airwicks were in great demand on these flights.

Handley Page were tasked to find a cure for the problem by determining if the environmental control of the Hermes could increase the cooling flow. The task was allotted in 1958 to the P.Jones of the Auxiliary Systems Section, and the investigation determined that the system's auto control setting of 45 lb/minute corresponded very well with the peak cooling flow for a good range of operating conditions. In other words the automatic setting had been nicely chosen for optimum cooling conditions for humans - no monkeys were mentioned in the report!

By 1959 some of the Hermes were on the move to other operators, or going on lease to foreign operators. G-ALDE and G-ALDL went to Bahama Airways as VP-BBO and VP-BBP respectively, G-ALDY went to Middle East Airways as OD-ACC, and G-ALDU and G-ALDX were leased to Kuwait. New or resurrected company names started to appear on some of the Hermes, and these carried on the use of the aircraft until the last one was scrapped in 1968. G-ALDK of Skyways finished its life at Nicosia, Cyprus, when EOKA terrorists blew it up on 4 March 1956.

Airwork lost another Hermes when G-AKFP landed on DC-3 VT-AUA, which was waiting to take-off at Calcutta, on 1 September 1957. George Piper:

"An Airwork Hermes, operated by a Skyways crew, was wrecked in a landing accident, also involving an I.A.C Dakota, at Calcutta. I was in Calcutta myself at the time awaiting the return of the aircraft from Singapore, and I accompanied my captain, Charles Stenner, who was also the Chief Pilot, out to Dum-Dum to inspect the wreckage. The aircraft was a complete write-off, but thanks to its rugged construction the occupants escaped with minor injuries."

At present at Duxford airfield there is the

Hermes IV G-ALDT of Skyways at London Airport. This aircraft, the fomer BOAC 'Hestia' was scrapped at Stansted in 1962. [MAP]

HERMES IN SERVICE

Hermes IV G-ALDU of Britavia at it's maintenace base of Blackbushe.

fuselage of G-ALDG, whose restoration was commenced under the project leadership of Peter Webster (crew chief Duxford), Harry Fraser Mitchell (at that time Chairman of the Handley Page Association), and W.E.(Bill) Bye (chief representative on the Hermes for the HPA). The work has been carried on by members of the Duxford Aviation Society and volunteers from the Handley Page Association.

In retrospect it can be said that the Hermes was a first-class attempt to produce an aircraft based on soundly based construction, but was, as with most contemporary aircraft of that time, only marginally adequately powered. It introduced improved pressurisation and environmental systems, and a high voltage electrical system, all of which brought 'teething' troubles. These problems, lack of decisions by Ministry and BOAC officials, and delays with the delivery of new equipment, meant a gestation period that was far too long for what had originally been envisaged as an 'interim airliner'. By the time the Hermes entered service with BOAC, the jet-powered airliner era was around the corner.

As an airliner the Hermes possessed a strongly built airframe that provided the passengers with an air-conditioning and environmental system as good as, if not better than, any installed in any American aircraft of the period. Passenger comfort was well catered for and the crew were provided with the most comfortable seats ever installed in aircraft. In its operations with the Independents its annual utilisation of 3,000 hours was on par with other piston-engined aircraft of that period. That it failed to garner any further orders was a disappointment to Handley Page personnel; but the effort expended in the development and design of equipment for the Hermes made easier the same process for the Victor bomber.

The Hermes was the last large civil transport aircraft built at Radlett, it was the last large British built, piston-engined airliner to go into service with BOAC, but as with all Handley Page aircraft it set a high standard of construction and a high standard of comfort. The Hermes airliner with its companion the Hasting military transport, were well-built aircraft - Handley Page aircraft.

Hermes IV G-ALDA of Falcon Airways [B. Pegden]

HANDLEY PAGE HASTINGS AND HERMES

Appendix One
Basic data on Hastings aircraft

	C1	*C2*
Maximum load	16,600lbs or 30 paratroops plus 20 containers.	20,311 lbs.
Basic equipped weight	50,600 lbs	-
Maximum take-off weight	75,000 lbs	80,000 lbs
Maximum landing weight	70,000 lbs	74,000 lbs
Wingspan	112 ft 9 in	112 ft 9 in
Wing area	1,408 sq ft	1,408 sq ft
Length, tail down	82 ft 9 in	82 ft 9 in
tail up	82 ft 1 in	82 ft 1 in
Tailplane span	43 ft	51 ft
Tailplane area	401 sq ft	442 sq ft
Track	24 ft 8 in	24 ft 8 in
Wing chord at root	16 ft 0.32 in	16 ft 0.32 in
at tip of outer wing	6 ft 11 in	6 ft 11 in
Controls: Elevator movement up	25°	25°
down	17°	15°
Aileron movement up	29°	29°
down	16°	16°
Rudder movement left	25°	25°
right	25°	25°
Flap movement fully down	80°	80°
Max lift	40°	40°
Undercarriage Main wheel	Dunlop AH9596 or 50519	Same as C1.
tyre	64 x 22.5 x 26in	
Tailwheel	Dunlop AH8291	Same as C1.
tyre	9.25 x 13 in	
Hydraulics Type	Electro-Hydraulics system	Same as C1
Power	2 x Lockheed Mk.6 EDPs.	Same as C1
Tank Capcity	4.5 gallons	Same as C1
System Capacity	7.75 gallons	Same as C1
Pressures	2800 psi pressure limit valve	Same as C1
	2450 psi cut-out valve	Same as C1
Engines Type	Hercules 101/106	Hercules 216
Fuel Capacity	C1 - 2563 gallons C1A - 3263 gallons	3175 gallons
Pneumatic system	Hymatic 600 psi	Hymatic 600 psi
Propeller Type	DH Hydromatic 4 blade DH/100/446/2 13ft dia.	DH Hydromatic 4 blade DH/100/446/4 13 ft dia.
Max speed at height	354 mph at 23,700ft	348 mph at 22,200 ft
Max weak mixture cruise	276 mph at 10,000ft	291 mph at 15,200 ft
Service ceiling	26,700 ft	26,500 ft
Range with max load	1220 miles	1690 miles
Max range	3260 miles	4250 miles
Payload for maximum range	5,250 lbs	7,400 lbs
Take-off over 50 ft barrier	1210 yds	1586 yds
Landing over 50 ft barrier	1375 yds	1430 yds

HANDLEY PAGE HASTINGS AND HERMES
Appendix Two
Basic histories of each Hastings aircraft

TE580	First protoype Hastings. AFEE 1946. A&AEE 1948
TE583	Second prototype. A&AEE 7.1.47. To AFEE for glider towing trials. 14.6.49. returned to Radlett for conversion to take two Sapphire jet turbines in outer nacelles. Used for flight testing of Victor type escape door. Used at Defford RRE for flight testing of numerous radio and radar equipment. 7.4.65 to Manston for fire fighting use.
TG499	First production a/c. A&AEE 1947. AFEE for paratrooping trials, trial installation of heavy dropping beam and Paratechnicon. On 26.9.49 Paratechnicon broke away from aircraft and aircraft destroyed at Beacon Hill, Wilts.
TG500	A&AEE. 27.4.47 A&AEE for trials. At AFEE for heavy dropping trials. For trials of micro-wave radar for MCA at A&AEE. Used at A&AEE for overseas ferry operations. SoC 12.4.73.
TG501	HP; A&AEE 16.1.47 for flight trials of equipment. To ETPS for test pilot training duties. RAE Farnborough for trials in North Africa and test of autopilot. SoC 29.6.66.
TG502	RAE; A&AEE 2.10.47 to RAE Farnborough for trials of radio and radar equipment. To A&AEE. for handling trials and also used to test lowered tailplane. Paratrooping trials at AFEE, and heavy dropping trials. SoC 29.6.66.
TG503	A&AEE 1947.; Fuel jettisoning trials at RAE. Take-off trials at A&AEE and AFEE. Demonstration flight to Australia and New Zealand. Performance trials at A&AEE. Flight tests at WIE Defford of radar equipment. RRE Reconnaissance Flight Upwood for initial trials of NBS equipment in Met I aircraft. Modified to Mk 5 aircraft. SCBS; Preserved RAF Gatow.
TG504	A&AEE ; 47; 202; SoC 8.8.66.
TG505	47; 202; 1.1.60 to A&AEE for trials as Mk T5.BCBS; SCBS; To SAS Hereford.
TG506	PTU; WEE; RAE; 28.9.48 to W.E.E for winterisation trials. Parachute development work at PTU and parachute test in North Africa. Henlow for

Images from the Hastings Sales brochure - 1
The Hastings was suitable for many roles; (1 and 2) - Ambulance; (3 and 6) Paratrooper; (4) freighter; (5) troop-transport; supplies dropping aircraft and glider-tug.

PRODUCTION

parachute testing. Then to A&AEE for overseas ferrying duties. 24-36; 202; SoC 4.10.66.

TG507 EFS; 47; 202; 24-36; 242 OCU; MoA; To Handley Page for special camera installation. RAF Wyton for special trials. SOC 23.4.69

TG508 EFS; 47; 202; 53-99; 242 OCU; crashed on landing 7.3.62

TG509 17.9.48 Arrived A&AEE for trials of gravity conveyor for supply dropping. 5.12.50 Arrived Defford for pattern aircraft for Hastings H2S-NBC installation. 53-99-511; 242 OCU; 70; SoC 9.11.67

TG510 47; 247 OCU; 24; 47; 1312 Flt; 511; 36; 242 OCU; SoC 1.11.66.

TG511 47; 99-511; 202; BCBS; SCBS; Preserved RAF Cosford as 8554M.

TG512 TCDU; 241 OCU; 53-99-511; SOC 27.11.59.

TG513 53-99; 511; 242 OCU; 70; 24-36; SoC 8.9.67.

TG514 47; Used at Handley Page for partial mock-up of VIP role. 3.10.50 for conversion to Met I role, 9.9.55 to RRE Defford for radio/radar installations and flight refuelling positioning equipment. SoC 1967.

TG515 TCDU; 53-99; 242 OCU; SoC 4.11.59.

TG516 53-99-511; 99; 36; 48; SoC 21.1.72.

TG517 47; 53; 202; SCBS; Preserved at Newark Air Museum.

TG518 47; 202; 53-99; BCBS; SCBS; SoC 13.5.69.

TG519 47; Crashed on approach to Dishforth. Became 6609M.

TG520 47; 24-47; 48; FECS; 48; FECS; 48; SoC 27.2.67.

TG521 47; CA; 53-99; 53; 242 OCU; BCBS; SCBS; SoC. 9.7.71.

TG522 47; 53-99; 36; Crashed on approach to Khartoum 29.5.59

TG523 47; 53-99; 47; 24; 70; 48; SoC 15.2.67.

TG524 47; 53-99-511; 114; A&AEE; 70; 24-36; 70; SoC 2.7.71.

TG525 47; 53-99; 48; SoC 9.3.67

TG526 47; 24; 24-47; 70; 48; SoC 9.3.67

TG527 47; 24-47; CA; 24; 24-36-114; 24; 16.5.60 to A&AEE for flight trials of UHF radio and radio compass. 23.3.64 to A&AEE for flight trials of Marconi VOR. BCBS; SCBS; SoC 4.11.68.

TG528 47; 53-99; 242 OCU; 24-36; 24; SoC 25.1.68; Sold to Skyframe Museum; To IWM Duxford.

TG529 47; 53-99; BCSS; SCSS; SoC 30.4.69.

TG530 47; 53-99; 242 OCU; 70; 151; 51; SoC 29.8.67.

TG531 53-99-511; 48; SoC 1.11.66.

TG532 53-99; 24; 36; 24-36; SoC

Images from the Hastings Sales brochure - 2
Military equipment up to a total of 7.5 tons is accomodated in the Hastings' fuselage. Seen inside the aircraft is (1) 17-pounder gun; (2) Bulldozer; (3) Scout-car (4 and 7) 40mm AA gun; (5) 3-ton lorry (6) 15 CWT truck. The large freight door (8) has a paratroop door incorporated into it.

HANDLEY PAGE HASTINGS AND HERMES

TG533 14.1.66
A&AEE. 22.11.48 to AFEE for trials of gravity roller conveyor equipment. Tropical glider towing trials and tropical towing trials of Hamilcar glider. 21.10.49 used as mock-up for conversion to Met 1 aircraft. 202; 242 OCU; 24-36; 70; SoC 18.12.67

TG534 Destroyed in ground fire at Dishforth 6.4.49

TG535 241-242 OCU; 70; 24-36; 24; SoC 13.1.68

TG536 241 OCU; 47; 47-53; 53-99-511; 242 OCU; 48; 242 OCU; BCBS/ SCBS; SoC as 8405M. Wings etc to Yorkshire Air Museum Halifax project.

TG537 24; 511; 36; 242 OCU; 24-36; 242 OCU; SoC 8.8.66

TG551 53-99; 242 OCU; 70; 24-36; 70; SoC 5.11.67.

TG552 Crashed at Lyneham during landing 12.4.51.

TG553 47; 99-511; BCBS; SCBS; SoC 1967

TG554 202; 242 OCU; 53-99; 242 OCU; SoC 4.11.59.

TG555 241 OCU; 242 OCU; SoC 6.2.60.

TG556 53-99-511; 24; 24-36; SoC 25.8.67.

TG557 53-99-511; 511; 36; 114; 70; 24-36; SoC 3.9.68.

TG558 241 OCU; 242 OCU; SoC 4.11.59.

TG559 24-47; Crashed on landing at Abingdon 9.10.53. became 7108M.

TG560 CSE; 116; SoC 4.3.58.

TG561 242 OCU; 70; SoC 8.2.67.

TG562 Crashed at Topcliffe on take-off 14.3.52.

TG563 241 OCU; 47; 53-99; 242 OCU; 99; 70; SoC 31.10.67

TG564 53-99; Crashed on undershoot on landing at Kai Tak 27.7.53.

TG565 Lyneham; 202; 242 OCU; 27.6.63 to A&AEE for flight trials of Ekco cloud and collision radar. 13.8.63 to 202 Squadron. 202; SoC 30.11.66.

TG566 241 OCU; Converted to Met 1; 242 OCU; 202; Crashed at Aldergrove on take-off 19.9.62.

TG567 241 OCU; 202; A&AEE; SoC 1.6.66

TG568 TCDU; PTS; 24; 53-99; 24; 19.5.59 to A&AEE for parachute clearance trials of a number of pieces of airborne support equipment. SCBS; 19.2.74 SoC on fire dump at RAE Bedford for fire fighting duties.

TG569 242 OCU; 53-99; 48; SoC 1.11.66

TG570 241 OCU; 242 OCU; 48; 242 OCU; SoC 11.11.67

TG571 241 OCU; 24; 1312 Flt; 99; 70; 24-36; 242 OCU; SoC 7.6.67

TG572 241 OCU; 242 OTU; 202; SoC 31.8.66

TG573 241 OCU; 47; 53-99; SoC 22.1.59. became 7594M

TG574 241 OCU; Crashed on approach at Benina 20.12.50 .

TG575 241 OCU; 24; 70; Crashed on landing at El Adem 4.5.66.

TG576 241 OCU; 202; 242 OCU; 70; 24-36; SoC 22.8.67

TG577 241 OCU; 242 OCU; 53-99; 511; 36; 70; 24-36; Crashed in circuit at Abingdon 6.7.65 due fatigue.

TG578 241 OCU; 53-99; 242 OCU; SoC 4.11.59

TG579 241 OCU; TCASF; 242 OCU; 48; Landed in sea on approach to Gan 1.3.60.

TG580 241 OCU; TCASF; 24-47; 48; Crashed landing at Gan 3.7.59

TG581 241 OCU; 242 OCU; 24-36; SoC 23.8.67

TG582 241-242 OCU; TCDF; 47; 24; 70; 24-36; SoC 30.12.65

TG583 241 OCU; Crashed on approach to Dishforth 31.7.50.

TG584 241 OCU; 53-99; 242 OCU; Crashed on overshoot 13.9.55.

TG585 241 OCU; SoC 8.2.60

TG586 241 OCU; 242 OCU; SoC 4.11.59

TG587 242 OCU; 57-99-511; 36; 242 OCU; SoC 17.8.67.

TG661 1 PTS; 24-47; 1312 Flt; 24; 47; 242 OCU; SoC 23.11.59.

TG602 1 PTS; TCASF; Crash after structural failure Shallufa 12.1.53.

HANDLEY PAGE HASTINGS AND HERMES

TG603 24-99; Blown off runway at Luqa. DBR 16.6.52.
TG604 241 OCU; 53-99; 242 OCU; 24-36; SoC 15.8.67.
TG605 24; 53-99; 24; 114; 24-36; SoC 7.11.67. became 7987M.
TG606 24-47; 24; 114; 70; SoC 30.11.66
TG607 24; 24-36-114; SoC 1.8.67.
TG608 24; 24-47; 511; 70; 242 OCU; SoC 21.9.67.
TG609 24; 242 OCU; SoC 8.2.60
TG610 53-99; 47; 53-99; 48; 242. OCU; Crashed landing at Thorney Island 17.12.63.
TG611 Dishforth/ Crashed on takeoff Tegel 16.7.49.
TG612 241 OCU; 242 OCU; 70; 48; SoC 14.11.66
TG613 47; 53-99; Ditched in Mediterranean 22.7.53.
TG614 24-47; TCASF; 24-47; 70; 48; SoC 9.3.67.
TG615 1 PTS; TCASF; 24-47; 1312 Flt; Crashed on approach Colerne 21.10.57
TG616 511; 53-99; 114; 36; 242 OCU; SoC 27.11.67.
TG617 RAFFC; 242 OCU; SoC 16.12.59
TG618 RAE: and Met Research Flt SoC 29.6.68
TG619 20.4.55 to RAE Farnborough for Met Research Flight, also used for flight trials of radio/radar equipment and warning radar. SoC .1.70
TG620 202; 36; 24-36; 48; SoC 9.3.67.
TG621 202; 70; 202; 24-36; 24; SoC 12.2.68.
TG622 16.11.56 to MoS Fleet at Radlett for flight trials of radio equipment. 202; SoC 31.10. 66
TG623 18.7.58 to A&AEE for flight trials of special-radar, also to RRE. 2.2.59 to Aldergrove Coastal Command. 202; SoC 3.3.67
TC624 202; Crashed on take-off Aldergrove 27.12.61
WD475 To A&AEE for handling trials, first C2. 24-47; 53-99-511; 511; 36; 114; 24-36; 70; SoC 6.12.67.
WD476 14.3.51 To A&AEE for flight trials at increased weight and take-off power, Paratrooping and supply dropping trials with Heavy Beam. Tropical trials and engine cooling trials at Khartoum. CA; 511; 24; 24-36; 24; SoC 25.9.69.
WD477 A&AEE; 511; 24-36; 24; SoC 30.1.68
WD478 RAFFC; Crashed on take-off Manby 19.3.51.
WD479 24-47; 24; 48; SoC 6.3.67.
WD480 1.4.53. To RAE Farnborough for sonobouy trials. Then used for radio, navigation and ILS trials. Aircraft reconditioned and fitted with Hercules 216 engines and used at RAE for other radio trials. SOC 25.9.74 Farnborough.
WD481 511; 53-99-511; 511; 36; 114; 48; SoC 25.4.67.
WD482 31.5.51 To RRE Defford for installation of H2S Mk 9. Used for ASV and other radar trials and radar equipment being developed for TSR2. SoC 1.5.67
WD483 24-47; 70 ; Crashed on landing @ Ataq 9.4.56
WD484 24.7.52 to RRE Defford for radio/radar trials. Take-off crash at Boscombe Down on 29.3.55 and aircraft destroyed.
WD485 24-47; TCASF; 1312 Flt; 99; 36; 114; 24-36; 24 SoC 25.9.69
WD486 1 PTS; 24-47; 511; 24-47; 24; 114; 70 SoC 29.11.67
WD487 1 PTS; 24-47; TCASF; 24; 24-36; 24 ; SoC 14.18.68.
WD488 24-47; 511; 53-99-511; 511; 36; 48 ; SoC 26.11.65.
WD489 47; 24; 70 SoC 12.2.68.
WD490 24-47; 48; 70 SoC 11.10.67 became 7985M
WD491 24-47; 53-99-511; 24; 24-36 crashed on take-off West Raynham 9.6.67.
WD492 47; crashed during supply drop in Greenland 16.9.52
WD493 TCDU; TCDF; 24; 24-36 SOC 2.2.68.
WD494 24; 47; RAFFC; 24-47; 24; 24-36; SoC 20.9.67.
WD495 47; 1312 Flt; 99; 36; 114; 24-36; 24 SoC 7.2.68.
WD496 CA; 27.3.53, to A&AEE for flight

HANDLEY PAGE HASTINGS AND HERMES

	trials of TI modifications and cabin temperature trials. Paratroop static line installation. Tests of modified roller conveyor installation Used in 1958 for flight tests across Atlantic and North America of Decca/Destra equipment. Used in support of Comet 1A with radio/radar equipment. Then used for a wide range of electronic, navigational and radar aids until SOC 22.11.72.
WD497	47; TCASF; 53-99-511; 511; 36; 48 crashed Seletar 29.5.61.
WD498	From 28.7.61 to 5.9.61 at A&AEE on TACAN radar trials. 24-47; TCASF; 48; 70 Crashed on take-off El Adem 10.10.61.
WD499	RAFFC; 53-99-511; 24; 24-36; 48; To MinTech; SoC 1967
WD500	A&AEE; 24; FECS; 24; 70 ; SoC 31.1.70.
WJ324	24; 70; FECS ; SoC 9.3.67
W5325	To MoS Fleet at Radlett for flight trials of IFF, ILS and Sperry equipment. FECS; 24; Station Flt. Khomaksar; MECS; FECS SoC 28.3.68
WJ326	24-47; FECS/ MECS SoC 2.7.71.
WJ327	RAFFC; 24; 99; 24; 24-36; 24; RAE Farnborough for radio laboratory work 6.11.67. SoC 29.5.73.
WJ328	511; 70; 36; 70. To A&AEE from 12.7.62 to 14.8.62 for flight checks on TACAN equipment. SoC 28.3.68.
WJ329	511; 53-99-511; 511; 36; 24-36-114; 24; SoC 25.9.69.
WJ330	RAFFC; 511; 99; 24; 114; 24-36/24; SoC 25.9.69.
WJ331	511; 99; 24; 114; 24-36-114; 24; SoC 15.2.71
WJ332	511; 53-99-511; 99; 24; 114; 48; FECS SoC 13.8.68.
WJ333	511; 53-99-511; 36; 48; FECS SoC 14.3.67.
WJ334	511; 36; 24-36; 24; SoC 5.2.68.
WJ335	511; crashed on take-off Abingdon 22.6.53.
WJ336	511; 48; SoC 5.11.68.
WJ337	511; 53-99-511; 70; 99; 24; 114; 48; FECS. To A&AEE for take-off measurements from 12.2.59 to 2.3.59. SoC 5.11.68.
WJ338	511; CSE; CSDS; 151; 97; 115. To RRE 21.1.69 to act as target at 15,00 to 20,000ft for ground radar. To Malaysia for trials of RN radar and same at RRE. Despatched to RAF Catterick 4.7.69. SoC 4.7.69.
WJ339	511; 53-99-511; 99; 24; 24-36; 24 SoC 25.9.69.
WJ340	517; 53-99-511; 24; 24-36; 24 SoC 7.2.68.
WJ341	511; 24-47; Crashed on landing Abingdon 26.7.55
WJ342	24-47; 47; 511; 36; DBR at Eastleigh Kenya 23.1.61
WJ343	511; 99; 36; 24-36; 24; SoC 25.9.65

Images from the Hastings Sales brochure - 3
(2) Hastings' large freight doors allow for a wide variety of equipment to be loaded usiing the air-transportable loading ramp. Being loaded is (1) 40mm AA gun; (3) bulldozer; (4) 15 cwt truck; (6) 17-pounder gun. The under-fuselage supply-dropping role is demonstrated in (5)

Appendix Three
Basic data for Hermes II and IV

	Hermes II	**Hermes IV**
Wingspan	113 ft	113 ft
Wing area	1,408 sq ft.	1,408 sq ft.
Wing chord at root	16 ft 0.32 in	16 ft 0.32 in
Length	95 ft 6 in	95 ft 10 in
Tailplane span	51 ft	51 ft.
Wing loading	54.68 lb/sq ft	58.24 lb/sq ft.
Total engine power	6,860 hp.	8,400 hp
Power loading	10,48 lb/hp	9.281 lb/hp.
Basic equipped weight	51,200 lb	55,350 lb
Maximum all-up-weight	80,000 lb	86,000 lb
Maximum landing weight	70,000 lb	75,000 lb
Payload	16,000 lb	17,000 lb
Maximum range	NK	3,080 miles
Service ceiling	25,100 ft	24,500 ft
Take-off distance to clear 50 ft	1,500 yards	1,362 yards
Initial rate of climb	1,200 ft/min	975 ft/min
Maximum speed	NK	355 mph
Maximum WM cruising speed	219 mph	276 mph
Engines	Hercules 120	Hercules 763
Propellers (de Havilland)	CD18/446/1	CD60/446/1
Total fuel load	2,563 gallons	3,172 gallons
Undercarriage	Messier	Messier
Undercarrriage track	24 ft 8 in	24 ft 8 in
Undercarriage wheelbase	48 ft	29 ft 8 in
Brakes	Dunlop	Dunlop
Wheels	Dunlop	Dunlop

NK = not known, as aircraft was only used as test vehicle

Appendix Four
Basic histories of each Hermes aircraft

Hermes I.
G-AGSS - HP68/1
Registered 2.12.45. Crashed on take-off on first flight at Radlett.

Hermes II
G-AGUB - HP74/1
Registered 2.9.48 to Ministry of Supply, Radlett. VX234 to RRE Defford October 1953. Scrapped 1968.

Hermes IV
G-AKFP - H.P.81/1,
Registered 14.10.49, for B.O.A.C. named *'Hamilcar'*; leased to Airwork Ltd., 1953, trooping serial XD632 in 1953; sold to Airwork Ltd. 2.57. DBR in take-off collision with Dakota VT-AUG at Calcutta 1.9.57.

G-ALDA - H.P.81/2.
Registered 14.10.52, for B.O.A.C. named *'Hecuba'*; leased to Airwork Ltd. 1952, trooping serial WZS38; sold to Airwork Ltd. 1.57; To Falcon Airways Ltd.10.59; Air Safaris Ltd., 12.60 To Air Links Ltd. 11.62; flown to Southend 22.12.64 and scrapped in 1965

G-ALDB - H.P.81/3,
Registered 30.11.49, for B.O.A.C. named *'Hebe'*; leased to Airwork Ltd. 1952, trooping serial WZ839; burned out in crash at Pithiviers, France 23.7.52.

G-ALDC - H.P.81/4,
Registered 14.12.49, for B.O.A.C. named *'Hermione'*; leased to Airwork Ltd. 1952, trooping serial WZ840; sold to Airwork Ltd. 1.57; To Falcon Airways Ltd.named *'James Robertson Justice'* 6.59; crashed landing at Southend 9. 10.60.

G-ALDD - H.P.81/5,
Registered 19.12.49, for B.O.A.C. named *"Horatius'*; To Skyways Ltd., 4.55. Withdrawn from use at C. of A. expiry 17.7.59 and scrapped at Stansted.

G-ALDE - H.P.81/6,
Registered 7.2.50. for B.O.A.C. named *'Hanno'*; To Skyways Ltd. 1.55; To Bahamas Airways Ltd. 1.60 as VP-BBQ restored. 5.61 to Air Safaris Ltd.; scrapped at Hum 5.62

G-ALDF - H.P.81/7,
Registered 27.2.50, for B.O.A.C. named *'Hadrian'*; leased to Airwork Ltd. 1952, trooping serial WZ841; lost at sea off Trapani, Sicily 25.8.52

G- ALDG - H.P.81/8,
Registered 9.3.50, for B.O.A.C. named *'Horsa'*; To Airwork Ltd. 8.57; To Falcon Airways Ltd. 10.59; To Britavia/Silver City Airways Ltd., named *'City of Chester'* 12.59; scrapped at Gatwick 10.62

G-ALDH - H.P.81/9,
Registered 20.3.50, for B.O.A.C. named *'Heracles'*; To Skyways Ltd., Bovingdon 8.55; DBR when undercarriage collapsed at Heathrow 8.3.60; scrapped at Stansted.

G-ALDI - H.P.81/10,
Registered 6.7.50, for B.O.A.C. named *'Hannibal'*; To Britavia Ltd.7.54, trooping serial XJ309; operated by Silver City Airways Ltd., named *'City of Coventry'*; flown to Stansted 10.10.62 and scrapped.

G-ALDJ - H.P.81/11,
Registered 7.7.50, for B.O.A.C. named

HANDLEY PAGE HASTINGS AND HERMES

For the Hermes Sales brochure, Handley Page had an artist create an impression of the aircraft in the cruise...

'Hengist'; To Britavia Ltd., 7.54. Destroyed in night landing crash at Blackbushe 5-6. 11.56.

G-ALDK - H.P.81/12,
Registered 12.7.50, for B.O.A.C. named 'Helena'; To Britavia Ltd., Blackbushe 7.54, trooping serial XJ281; DBR in crash landing at Drigh Road, Karachi 5.8.56.

G-ALDL - H.P.81/13,
Registered 21.2.51, for B.O.A.C. named 'Hector'; To Skyways Ltd. 6.55; To Bahamas Airways Ltd. 1.60 as VP-BBP; restored. 4.61 to Air Safaris Ltd.; To Skyways Ltd. 12.61; To Air Links Ltd. 8.62; flown to Southend 31.8.62 and reduced to spares for G-ALDA.

G-ALDM - H.P.81/14,
Registered 17.7.50, for B.O.A.C. named 'Hero'; To Air Safaris Ltd. 11.56, leased to Silver City Airways Ltd. until 12.59; scrapped Hurn 5.68.

G-ALDN - H.P.81/15,
Registered 20.7.50, for B.O.A.C. named 'Horus'; forced landed in the Sahara Desert 150 miles south east of Port Etienne, French West Africa 26.5.52.

G-ALDO - H.P.81/16,
Registered 20.7.50, for B.O.A.C. named 'Heron'; leased to Airwork Ltd. 1952; scrapped at Blackbushe 3.59.

G-ALDP - H.P.81/17,
Registered 24.8.50, for B.O.A.C. named 'Homer'; To Britavia Ltd. 7.54, trooping serial XJ269, operated by Silver City Airways Ltd. named 'City of Truro'; flown to Stansted 10. 10.62 and scrapped.

G-ALDR - H.P.81/18,
Registered 29.8.50 for B.O.A.C. named 'Herodotus'; To Skyways Ltd. 4.55; withdrawn from use at C. of A. expiry 8.59 and scrapped at Stansted.

G-ALDS - H.P.81/19,
Registered 6.9.50, for B.O.A.C. named 'Hesperides'; To Skyways Ltd. 4.55; withdrawn from use at C. of A. expiry 1.60 and scrapped at Stansted.

G-ALDT - H.P.81/20,
Registered 13.9.50, for B.O.A.C.named 'Hestia'; To Skyways Ltd. 4.55 to M.E.A. 6.55 as OD-ACB; restored. 10.55; to Bahamas Airways Ltd. 10.60 as VP-BBQ restored. 6.61 to Air Safaris

HANDLEY PAGE HASTINGS AND HERMES

...and landing configurations. The nosewheel undercarriage is noticable in this biew.

Ltd.; To Skyways Ltd.12.61; withdrawn from use at C. of A. expiry 6.62 and scrapped at Stansted.

G-ALDU - H.P.81/21,
Registered 12.10.50, for B.O.A.C.named 'Halcyone'; To Britavia Ltd. 7.54, trooping serial XJ280; To Kuwait Airways Ltd. 7.56-1.57; operated by Silver City Airways Ltd. named 'City of Gloucester'; scrapped at Stansted 1 1.62.

G-ALDV - H.P.81/22,
Registered 29.9.50, for B.O.A.C.named 'Hera'; To Skyways Ltd.4.55; burned out in test flight crash at Meesden Green, Herts. 1.4.58.

G-ALDW - H.P.81/23,
Registered 30.10.50, for B.O.A.C. named Helios" To Skyways Ltd. 4.55. Blown up by saboteur at Nicosia, Cyprus 4.3.56.

G-ALDX - H.P.81/24,
Registered 13.12.50, for B.O.A.C. named 'Hyperion"; To Britavia Ltd. 7.54, trooping serial X5267; To Kuwait Airways Ltd. 7.56 to 1.57 w.fu. 1.60

G-ALDY - H.P.81/25,
Registered 16.1.51, for B.O.A.C. maned 'Honor'; To Skyways Ltd. 9.54; to M.E.A. 6.55 as OD-ACC; restored. 10.55 withdrawn from use at Stansted 12.58.

Handley Page H.P.82 Hermes 5
G-ALEU - H.P.82/1,
Registered 14. 10.48 to Handley Page Ltd., Radlett; DBR. in wheels- up landing at Chilbolton 10.4.51;to spares at Boscombe Down. Sold to Airwork Ltd.

G-ALEV - H.P.82/2,
Registered 14.10.48 to Handley Page Ltd., Radlett; dismantled at Farnborough 9.53 fuselage by road to Southend 1958 for use by Aviation Traders Ltd. for mock-up of freight door installation.

Handley Page Association

The Association was formed in 1979 to keep alive the memories of the Handley Page Companies, their Founder and their aircraft, and to foster the spirit of innovation, engineering ingenuity and excellence of construction that were hallmarks of the Company. Individual or Corporate membership is open to all those with an interest in the Company and their achievements.

Informative Newsletters are published regularly, giving news of people, aircraft and events relating to Handley Page and to the Association. Other activities include re-unions, talks, film shows, visits to museums and other places of interest, and social events, including a very popular "Night of Nostalgia". There is an active interest in the display of Handley Page aircraft and artifacts.

Photographic and documentary Archives are maintained for researchers and lecturers. Of particular interest in the current context, HPA volunteers are working to preserve and present for public inspection, the last remaining Hermes fuselage, now at Duxford Airfield, Cambridgeshire.

The Association is a Member of the British Aviation Preservation Council, and has links with the Yorkshire Air Museum, the Halifax Association, Canada, the 35 and 635 Squadrons Re-Union Association and the Croydon Airport Society.

Our Patron is Mrs Anne Manley Walker, eldest daughter of Sir Frederick.

The Author is an Hon. Life Member of the Association.

For further details please write (with SAE please) to:-
The Hon. Membership Secretary, B.Bowen,
77 Bowershott, Letchworth, Herts SG6 2EU U.K.

Aero Book Company

QUALITY BOOKS - QUALITY SERVICE

AVIATION BOOKS BY POST
(WORLDWIDE MAIL ORDER)

CATALOGUES AVAILABLE ON REQUEST

Aero Book Company, PO Box 1045, Storrington,
Pulborough, West Sussex RH20 3YD
Telephone 01798 812222 Fax 01798 813655